Hydroponics

A Complete Guide to Start Your Own Hydroponic Garden

Mark Greenwood

Table of Contents

1. Intruduction

Hydroponics is a method of growing plants without having to plant them in soil. Instead of soil, plants are planted in mineral nutrient solutions that are soluble in water. They may also be grown in gravel or perlite which are called inert mediums.

Hydroponics is not a new method of growing plants and can be traced back to ancient times. Examples of these are the floating gardens of the ancient Aztecs, the Hanging Gardens of Babylon, and the writing of Marco Polo indicate he saw similar gardens in China during the 13th century.

Hydroponic gardens were used to feed troops stationed on arid islands during various wars and have been introduced into space programs. Although the basic concepts of hydroponic growing are still the same, the methods have advanced in leaps and bounds over the past century.

The Basics of Hydroponics

Simply put hydroponics is a way of growing plants without soil. Instead, their roots are submerged in water and their needed nutrients are provided by a water-soluble medium. The word hydroponics is derived from the Greek words' "hydro" meaning water and "ponos" which means labor or working - "water working". Hydroponics is a way for people to enjoy freshly grown food in places where the soil is not suitable for traditional farming. This means it's perfect for people in places like the Sahara Desert or overpopulated urban areas where land is scarce. One such place is Bermuda where land is really scarce; they are now using hydroponics to ensure year-round fresh produce.

60-30 BCE - Hanging Gardens of Babylon

1110 - 1522 CE - Floating Gardens of the Aztecs of Mexico - Chinampas

1600 - Jan Baptist van Helmont - Experiment proved that plants obtain substances from water.

1860 - 1861 – Knops and Sachs - Plants can be grown in an inert medium moistened with a water solution containing nutrients - nutriculture.

1929 - Solution culture can be used for large scale crop production.
1937 - The term Hydroponics has been introduced.

Then there are very cold regions, like Alaska, that have a limited growing season, so they have begun using hydroponic greenhouses.

Hydroponic systems tend to have a plant growth rate that is twenty percent faster than plants grown in soil. The yields are also twenty percent larger than their soil-grown counterparts. Another advantage is that they save on water as the plants only draw what they need from the reservoirs. Some Hydroponic growing methods have a recovery system for reusing nutrients that overflow in excess.

Plants in a hydroponic garden does not have to grow large root systems in order to hunt for nutrients as it is all supplied by the growing medium. This means that plants can be placed side by side to save on space, making hydroponic growing an excellent solution for people who live in apartment blocks.

There are six different hydroponic systems to choose from; each one has its own advantages, but the systems recommended for beginners are the "**ebb and flow**", "**water culture**" and "**wick**" systems. These three systems can be bought in kits that are easy to get started with and use.

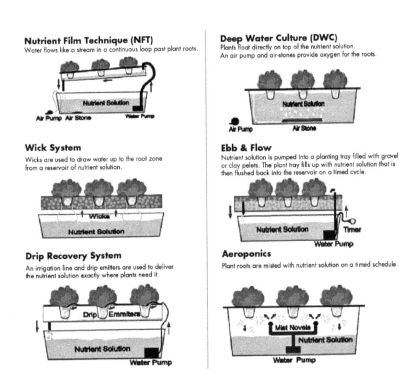

Nutrient Film Technique (NFT)
Water flows like a stream in a continuous loop past plant roots.

Deep Water Culture (DWC)
Plants float directly on top of the nutrient solution. An air pump and air-stones provide oxygen for the roots.

Wick System
Wicks are used to draw water up to the root zone from a reservoir of nutrient solution.

Ebb & Flow
Nutrient solution is pumped into a planting tray filled with gravel or clay pelets. The plant tray fills up with nutrient solution that is then flushed back into the reservoir on a timed cycle.

Drip Recovery System
An irrigation line and drip emitters are used to deliver the nutrient solution exactly where plants need it.

Aeroponics
Plant roots are misted with nutrient solution on a timed schedule.

There is a lot you can grow hydroponically like herbs, vegetables, varieties of flowers, and even some fruit. Since the pH levels can be controlled a lot of fruit that need various specific soil conditions, such as blueberries, do well in a hydroponic environment. Some plants like potatoes, radishes, and other big root vegetables may take a bit more practice to grow in a soilless system. But if you can get their environment right, even these plants outdo their soil planted counterparts.

2.   What is a Hydroponic System?

The term Hydroponics derives from two Greek words – 'hydro' meaning water and 'ponos' meaning labor. Hydroponics is the art of growing plants without soil. Most farmers in ancient times believed that soil provided only support to plants and it was water that helped plants grow – this was because their crops perished if the rains failed.

Apart from providing support to plants, soil is also a breeding place for millions of microbes – both aerobic and anaerobic. Aerobic microbes are those that help in the growth of plants; anaerobic microbes are those that cause plant decay.

Using a Hydroponic system, we can eliminate the presence of anaerobic microbes; the contribution normally made by aerobic microbes is now performed by supplying nutrient solutions.

A system that provides the essential elements for successful plant growth, without using soil as a growth material, is called a Hydroponic system. The essential elements are:

- Nutrient solution
- Proper exchange of gases
- Lighting

- Growth / Support material like Perlite, Rockwool, coconut coir, and so on.

The ancient Babylonians, the Aztecs and the Egyptians are said to have used Hydroponic irrigation to grow plants; the famous 'Hanging Gardens of Babylon' is one such example. Experiments in Hydroponics started in 1936 – Dr. Gericke from University of California, Los Angeles, successfully grew tomatoes in water culture. During World War II, Hydroponics was used to grow plants in non-arable areas. However, it was not until the 1970s when farmers and gardeners started showing interest in Hydroponics. But, even to this date, the potential of Hydroponics is not used to its fullest; many gardeners are still unfamiliar with the terminology.

Currently, Hydroponic systems are used to grow a variety of plants such as lettuce, green pepper, lettuce, basil, and so on. The Hydroponic system of cultivation is seen as a possible solution to the hunger problems of the world. The beauty of this system lies in its simplicity. You can use your garage, rooftop or any available space to grow plants. Continual research and development has led to the use PVC material that is reusable and long lasting; Middle Eastern countries, where water is a scarce resource, are adopting Hydroponic system for cultivation. Huge buildings with desalination systems are built to help in crop cultivation. The future of Hydroponic cultivation is getting brighter by the day and it would not be a surprise to see this method of cultivation becoming the norm!

Hydroponics gardens have quickly grown in popularity among consumers and restaurants as a fresh alternative to store bought produce. If you've ever tasted a fresh, naturally grown tomato you're familiar with how much better tasting they are than the average tomato from your grocery store. That's because traditional farm grown produce is typically picked prematurely and artificially ripened during transport. Plus, most of the produce found in your local supermarket is generally bred for hardiness and appearance, rather than taste.

With all of the advantages of hydroponics over traditional soil based gardening its not hard to see why so many people are starting their own hydroponics gardens!

- Use up to 2/3 less water
- Provide higher yield per square foot
- Perfect for city dwellers and those that don't have backyards
- Eliminate damage from pesticides and other harmful chemicals
- No digging, weeding, or back breaking work
- Fresh produce can be grown year round

Hydroponics may seem a little complicated at first but once you learn the basics it really gets quite easy and fun. There are a couple of main factors that will contribute to the success of your hydroponics garden.

3.   Hydroponic Gardening

What you are about to create, with the help of this hydroponic guide, is a method of growing vegetables, fruits, and herbs, without the use of any soil.

The roots will rely on a nutritionally enriched liquid. They are nourished in water along with liquid nutrients. You can use other mediums, such as perlite

and vermiculite, or even rockwool or clay pellets, but more of that later.

There are various ways of running this type of garden. In this book, I will explain each method to help you decide which one works best for your personal situation. Then, you will be ready to buy a kit, or put together your own equipment, according to the method you choose. To do so, you need to consider the following:

What is the size of the space for your hydroponics setup?

If you are new to hydroponic farming, then it is better to start small.

Is it indoors or outside?

If you are fortunate enough to have a reasonably sized yard, then you should consider setting up a greenhouse system. Hydroponics can be set up outside but you might be more prone to pests or the vagaries of the weather. If you have a spare indoor room, you will need to consider the light source.

Cost?

You will need certain types of equipment but you can limit the cost if you

start small.

What type of plants do you wish to grow?

You can grow just about any plant using hydroponic growing, such as vegetables, fruits, salads, herbs and even flowers.

What time do you have available for maintenance?

If you are a novice, it is better to grow fast-growing plants. There are plenty to choose from, such as lettuce, parsley, tomatoes or strawberries. None of

these need much in the way of your time. This way you learn at a steady pace and can change your system, or plants, as you become more experienced. If you enjoy it, then you can move on to the more complex plants.

PROS/CONS OF HYDROPONIC GARDENING

Well, for one thing, you won't get dirt in your nails, so that's a good start!

One of the most outstanding features of growing using this method is that plants use much less water than they would if grown as crops in the ground.

Hydroponic farming uses only 10% of the water that traditional ground crops use. Now that's a figure worth considering.

Could this be the future of farming in countries where water is a scarce resource? Already it is growing more popular. So, let's look at the positive features of a hydroponic garden, or farming system:

PROS

No Soil, Less Land

No soil whatsoever. Your plants will grow in a water-based system. Liquid nutrients will help them grow to maturity. The vast acres of land now used for farming can be used for other needs such as housing and forests. It also means that more vegetation can be grown in smaller plots.

Less Water

You can grow plants anywhere, anytime of the year, regardless of climate.

The system you choose will only use a set amount of water according to the size of your farming system. With the aid of simple equipment, the plant roots sit in the water. This is unlike field crops whereby the water either soaks off into the ground or dries up with the heat of the sun. The water in your system can be re-used time and again, so you are recycling it. No irrigation of the land is needed, thus fewer costs for the farmer.

Fewer Nutrients, Less Fertilizer

Nutrients are fed to the plants in a controlled environment instead of running off and soaking into the ground, polluting land and rivers. Imagine a farming world without the need of spraying fertilizer all over the land.

More Crops, Fewer Enemies

The indoor systems, such as inside your home or a greenhouse, have good advantages. There is no loss of crops due to bad weather. Wild animals cannot eat the plants as they can with field crops. With no soil, there will also be fewer pests and diseases to contend with.

The outdoor systems still work efficiently but if your hydroponic garden is not protected, then it could still be prone to problems from pests and the weather.

Healthier Crops

Crops grown by the hydroponic method enjoy 100% of the nutrients fed to them. None will soak into the ground or be blown away by the wind. The result of this is that they produce up to 30% more foliage than soil grown plants and grow 25% faster. It is all down to the well-balanced nutrients the roots receive from the water.

Fewer Chemicals Needed

With fewer parasitic bugs, fewer insecticides and herbicides are needed. The result is healthier food for human consumption.

Weather Resistant

Hydroponic crops grown indoors are not weather dependent. They can be grown all year round, regardless of the climate or temperature. Crops are

more protected if grown indoors or in greenhouses.

Less Labor

There will be no labor-intensive weeding necessary, either chemically, or by hand. This lowers the need for maintenance. There will be the initial seeding,

feeding, and harvesting which are achieved with much less labor than the traditional methods of growing crops.

Can be Grown Anywhere, Geographically

This method does not rely on available land. The farm could be set up near to the market where the crops will be sold. It's a great way of cutting down on transport costs and pollution. It can even be mobile, if necessary, and set up wherever and whenever needed.

A Method to Suit all Budgets

There are many methods to growing crops using a hydroponic system. It can be a small-scale affair in your backyard or indoors in a spare room. But, it can also be done on an industrial scale. Large volumes of produce with thousands of plants for a nation of people can be achieved using hydroponic farming.

CONS

Plants are dependent on humans. Nature has little to do with this type of farming. The plants are relying on human attention for everything, from water to food, to light and humidity. Once the garden is up and running, it can go into automation for the most part. Though as with all farming, someone must regulate it. If it is not done correctly, whole crops can be lost.

Requires some Expertise

This is not a traditional method of growing plants. A certain amount of knowledge on the various systems is required. Done incorrectly, the whole crop of plants could perish.

Safety

Everyone is aware that water and electricity are a dangerous combination.

Hydroponic plants need both water and electricity to manage the entire system. If mistakes happen, it could cause a life-threatening situation.

Electric Failure

What would happen if the electricity supply was down for any amount of time? This must be considered from the onset. If such a thing were to occur, without emergency provision, the whole crop could die in hours.

Initial Outlay for Large Farms

There is a need to buy equipment when you first set up your hydroponic system. That can be costly if farming on a large scale. Once established, the running costs will be electricity, water, and nutrients. Plus, a small labor force if it is a large farm.

Fast Spreading Diseases

The chances of soil disease are nil, so pests and diseases are fewer. However, if your system gets a water-based disease, this will spread rapidly to any plants on the same system. A means of measuring the water for such bacteria needs to be put in place to avoid this happening. Otherwise, you could lose your entire crop to disease.

4.   How Does a Hydroponic System Work?

A Hydroponic system must satisfy three basic requirements for plants to grow and survive:

Efficient supply of water and nutrients

Protection for plant roots from dehydration in the event of a pump or power failure

Maintenance of proper gas exchange levels between roots and nutrient solution

Schematic Representation of Hydroponic System:

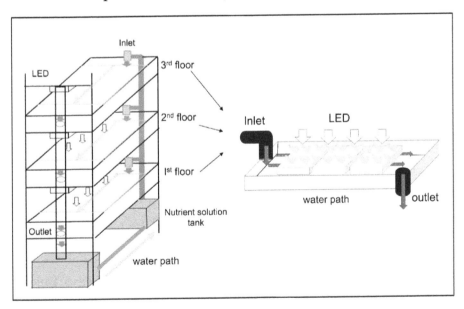

Seed selection has a significant effect when deciding on the taste, flavor, color and texture of your produce.

If your neighbor grows tasty and juicy tomatoes in the backyard, your best bet is to get seeds from their plants to get the same results. When plants grow outdoors, nature introduces genetic material that alters the result. When you grow plants indoors, you can introduce the same genetic material to grow produce according to your choice. You can alter or retain the same genetic coding in your next generation plants to maintain good quality.

Soil material in Hydroponics is replaced with loose growth / support material to drain nutrient solutions quickly. If you have a system that recirculates nutrient solutions frequently, use a material that drains the solution quickly. If you have a system where recirculation of nutrient solution is minimal, use a slow draining material. Apart from efficient draining, maintaining a good nutrient to air ratio is important. You can combine various materials to achieve high draining efficiency while maintaining a good nutrient to air ratio.

Based on your nutrient circulation method, you can either opt for an active system or passive system. An active system uses mechanical pumps to circulate nutrient solution and air; you can use timers to water your plants. A passive system is a closed system Choose the correct growth material

Select an Active / Passive to circulate nutrient solution Select appropriate lighting system Protect plants from pests, algae, fungi where nutrient solution is not recirculated. You can buy nutrient solutions or prepare it at home. When you prepare nutrient solution, care must be taken to use an accurate measuring system, and check and balance the pH frequently.

Plants require light for photosynthesis – a process that provides food for plants to grow.

In outdoor gardening, this light is provided by the Sun. With indoor gardening, you must provide alternate sources to help photosynthesis in your plants. This can be provided using HID – High Intensity Discharge lighting. HID lamps provide maximum PAR (Photosynthetically Active Radiation) for the amount of power consumed. As a rule of thumb, 20 to 50 watts lighting is required for every square foot.

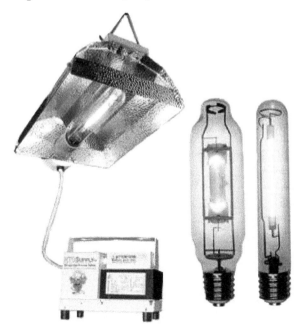

An indoor garden is free of many pesticides that are used outdoors. Also natural elements, such as rain, water and wind that keep the outdoor plants pest free, are absent in indoor gardening. However, one of the easiest ways to combat pests in your indoor garden is to keep it clean from dust and debris. Avoid going near your plants with unclean slippers. Keep a separate toolbox needed for your gardening and always remember to clean your tools with 10% bleach solution. Molds and mildew thrive in damp and humid atmosphere. Therefore, good air circulation is required to remove excess humidity. Maintaining excellent air quality is important to stop spores in the atmosphere from forming fungi. The indoor growing area must be clear of dust and debris. Maintaining correct watering levels is vital with indoor gardening. Humidity must be maintained between 60% to 80% and air must be continually circulated in the garden.

5.Growing Conditions

Whether your hydroponics garden is kept indoors or outdoors, a carefully controlled environment is essential to your garden's success. The three main factors to giving your garden a good home are humidity, temperature, and air circulation.

Water and Nutrients Solution

Starting with good water is crucial to any successful hydroponics garden. With traditional soil-based gardening the soil acts as a buffer and filter for any impurities that might harm your plants. In hydroponic gardening this filter is removed from the growing process so impure water or imprecise nutrient mixtures will have a direct negative impact on your hydroponics garden.

Lighting

One of the basic elements that plants need to grow. Assuming you're keeping your hydroponics garden indoors, you'll have to make sure you give your plants the necessary light they require to grow. This can be either a big, sunny windowsill or specialized plant lighting such as fluorescent, HID, or LED. The type of plants you end up growing will be a major factor of how much and what type of light you'll need.

Propagation

You'll have to decide if you're going to start your hydroponic garden using seeds or seedlings. If starting from seeds, which is the preferred option, you'll need a way to sprout them. There are several methods including using rockwool starter cubes, the most popular and common method, to help incubate your seedlings.

Medium

Since you're growing without soil, you'll need something else to hold your plant's root system together. The best growing mediums are chemically inert so they won't harm your plants or effect your nutrient solution, are inorganic so they won't breakdown and pollute your solution, porous enough to allow sufficient oxygen and water to reach your plants, and large enough so it doesn't fall through the slits in your net pots. There are several popular options that will all give your plants the proper foundation for their roots to hold on to including hydroton, coco, and silica.

Measuring Equipment

The pH of your water will be a major factor of the success of your hydroponics garden. Most plants will only grow in a pH range of

5.5 - 6.5 so you'll need a way to accurately measure it and adjust it if necessary. Luckily, decent pH meters are fairly inexpensive as are the appropriate pH up and pH down solutions. A combination thermometer/hygrometer is a nice-to-have tool to make sure your temperature and humidity levels are in the optimum range. Electrical timers are a great way of making sure your plants are getting the optimal amount of light each day.

Replicating Mother Nature and giving your plants ideal growing conditions sounds like a big task but its much easier than it sounds. Plants want to grow and will grow if given the most basic of conditions - just like this plant growing out of a crack on a rooftop of all places! It's up to you to give your plants the most optimal of conditions possible so they can provide you with the most optimal harvest possible. Of course, the larger your hydroponics garden the harder this task becomes but with the proper equipment and planning there are no limits to what you can grow.

Humidity

Anyone who has ever been in a greenhouse would think they need to crank up the humidity for their hydroponics garden. With traditional soil based gardening this may work best but your soilless garden is a bit different. High humidity will actually suffocate your plants and create a breeding ground for mold so you'll want to monitor your relative humidity levels. To maximize nutrient intake and promote optimal vegetative growth and flowering your hydroponics garden should be kept at a relative humidity range of about 40-80% with 50% being optimal.

A great tool to pick up is a combination hygrometer/thermometer that will allow you to keep tabs on your relative humidity and temperature. If you live in an area with naturally high or low humidity levels, or if you plan to keep your garden in an area of your house with high or low humidity levels, picking up a humidifier or dehumidifier may be a good investment.

Temperature

Plants are very sensitive to temperature and you must avoid extreme heat or extreme cold at all costs. The optimal temperature range for most plants is between 65° - 75°F. Sounds easy enough since you're growing indoors and this is the same range that most people enjoy but depending on the size of your garden, where you keep your garden, and what type of lighting you use, this may or may not be so simple.

If you only have a couple of plants and have a nice sunny window that will provide ample direct sunlight you can skip the rest of this. For those that have bigger things in mind though - keep reading! For larger gardens heat is going to be your #1 problem since you'll need supplemental plant lights and some of these lights produce a ton of heat. If this sounds like you here are a couple of tips to help keep your temperatures down:

- Make sure your garden has plenty of circulation and ventilation to vent away heat
- Open a window to allow cool air in and hot air out
- If needed consider a portable A/C unit
- Use grow lights with exhaust tubing and fan to vent away heat

Another thing to keep in mind is that there are warm season crops and cool season crops. Warm season crops such as tomatoes, peppers, and herbs enjoy a temperature range between 70°-80°F. Cold season crops on the other hand, such as lettuce, grow best at temperatures between 60°-70°F.

Saving the best tip for last - plants are hardwired and have evolved to adjust for a natural drop in nighttime temperature of about 10°F. This nighttime period of darkness and coolness is basically when plants sleep and, like all living creatures, this sleep period is vital to help them function properly.

Air Circulation

Proper air circulation is a must for any hydroponics garden for a couple of reasons. Plants rapidly deplete the air of carbon dioxide so fresh air (and carbon dioxide) is a must. Air circulation is also vital in order to vent away any hot, stale air that may suffocate your plants. Lastly, a nice gentle breeze will do wonders to help stimulate growth.

If you're growing a couple of plants in an open area with natural sunlight or using lights that don't give off a ton of heat such as LEDs or Fluorescents, a basic standing floor fan set on low will do the trick. If you're growing in a closed space such as a dedicated room or closet using hot HIDs though you may want to look at more powerful methods depending on how hot your room gets. If you're able to keep the temperature moderate enough an open window and floor fan should be sufficient; otherwise you may have to invest in an exhaust fan similar to a bathroom vent fan.

Water and Nutrients Solution

Now that you know what kind of environment you'll need to have for your hydroponics garden its time to look at the basic elements that will make your garden thrive. The foundation of your garden is going to be the water you use - similar to gas and cars, the water you use in your hydroponic system is what will make it really go! Depending on the mineral content of your water, you may or may not be able to use the water straight from your tap as your starting point. Since there is no soil to act as a filter for your plants they will absorb whatever is in the water you give them - if the water you start with has a high ppm count (parts per million) they'll absorb all these mineral particles as well as the nutrients you give them. Worse yet, nutrient/mineral lockout may occur and they may not even get the nutrients you intend to give them!

Having your tap water checked is pretty straightforward and cheap. You can sometimes find free test kits at your local hardware store that you can mail in or you might even be able to call your water company and ask. If you really want to get crazy they also sell ppm test kits and tools that allow you to measure this yourself. If your ppm is above 300 your best bet is to use water that has been treated with Reverse Osmosis filtration. You can install an RO filter kit in your house for around $100 or just find a local water store as they usually filter their water through RO. Distilled bottled water also makes a great option but will be the most expensive since you'll be paying about $1 per gallon. If you do end up trying to use regular tap water make sure you let it sit in the sun for at least 24 hours to let any minerals settle and dissipate.

pH

Hydroponics is a more scientifically based method of growing plants than traditional soil-based gardening and, unfortunately, it's time to bring things back to high school chemistry. pH, or potential hydrogen, is the measure of the activity of the hydrogen ion and acts as an indicator of alkalinity and acidity. The reason this is important in hydroponics is that plants require a pH range of 5.5 - 6.5 in order to grow with an optimal pH of 6.0. Anything lower will result in your solution being to acidic and anything higher will be to alkaline and prevent nutrient absorption.

Pure water has a natural pH of 7.0 so you'll need to lower the pH of your water. This is easily done using a couple of drops of a pH Down solution per gallon of water (or pH Up if the pH is on the low side). RO water has a pH right at 7.0 but there is some debate over the pH of distilled water. Many say that it has the pH of natural water at 7.0 but there are others who have measured it closer to 5.5.

Either way, a pH meter is a must have so you can always make sure you measure the pH of your water yourself and measure it often.

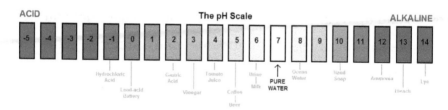

Temperature

Simple, easy to forgot, yet very crucial is the temperature of the water you use. If you're keeping your ambient temperature in the optimal range this shouldn't be a problem but you should still check in on this every now and then especially in the cold winter months and hot summer months. Ideally you'll want to keep your water between 65°-75°F. Any colder will freeze your plant roots and any higher will cook them as well as promote algae.

Nutrients

Part of the magic that makes it all happen! Plants require specific concentrations of three main macro elements and a number of trace micro nutrients in order to grow. Normally with traditional soil based gardening plants get all the nutrients they need straight from the soil they're planted in. In hydroponics we remove the soil so its up to you to provide your plants with the nutrients they need.

On most packages of nutrients and fertilizer you'll see three letters and associated numbers: N, P, and K. The letters and numbers represent the three macro nutrients and their ratios: Nitrogen, Phosphorus, and Potassium. You'll often find a wide range of NPK values as this allows you to easily adjust the ratio of each macro nutrient based on your plants' growth stage as each stage requires a different ratio for optimal growth. For instance, during the vegetative and growth stages the "N" number will be higher while in the flowering stage the "P" will be higher. Don't be too concerned with the trace elements such as sulfur, chlorine, zinc, iron, boron, and copper as these are all included in their proper ratios as well.

Nutrient Solution

While one of the main advantages of hydroponic gardening is simplicity, picking the proper nutrients can be an overwhelming task for both experts and beginners alike. The seemingly never ending choices of nutrients, additives, boosters, fertilizers, and other assorted potions are enough to make your head spin if you start over thinking things.

The trick here is to follow a good nutrient schedule and to keep things simple as most all purpose nutrients will provide all your essential elements needed for proper growth. Try to stay away from using too many different products as this will make it extremely difficult to troubleshoot potential nutrient issues. If you're not sure where to start General Hydroponics FloraSeries is one of the most widely used and trusted nutrient mixtures on the market today. Commonly regarded as the industry standard, this is the same solution that NASA chose to use in their hydroponics research.

When mixing your nutrient solution you should always remember to check the pH of your water first to make sure its right around 6.0. Stir in your nutrients, wait about an hour for things to settle, then recheck your pH to make sure its still in the optimal range and adjust if necessary. Some nutrient solutions may come in multiple parts such as the FloraSeries. Make sure to follow the instructions on which to add first and always mix well after adding each part as incorrect steps may result in nutrient lockout. Last, but not least, finish things off by adding a couple drops of hydrogen peroxide to your nutrient solution to help inhibit bacteria and algae growth (just make sure to double check your pH).

Lighting

If you thought nutrients were confusing, indoor plant lighting takes things to a whole new level. From fluorescents to HIDs to LEDs and ballasts and timers and hoods, lighting can be quite intimidating if you let it be.

K.I.S.S

The most important thing to remember about lighting for your hydroponics garden is to keep it simple. Most people have grown common indoor plants without a second thought about lighting. Depending on what type of produce you're looking to grow with your hydroponics garden you might be able to get away with using this same type of basic approach. Simple leafy vegetables such as spinach or lettuce and most types of herbs can be grown without spending a small fortune on specialized grow lights.

That's not to say you can throw your hydroponics garden in a dark corner of a spare room and think you're going to produce results though. A bright, sunny window with access to at least 5 hours of direct, natural sunlight and another 10 hours of indirect light will be more than enough to keep your leafy vegetables nice and happy.

If you notice your seedlings growing very tall and spindly this could be a sign they're not getting enough light. The reason for growth like this is that your seedlings are stretching themselves trying to get closer to your light source. If this is the case you may need to invest in some supplemental lighting.

Grow Light Systems

While its possible to grow most leafy vegetables using nothing but lots of direct, natural sunlight supplemental plant lighting is a great way to super charge your garden and is a must to grow plump, tasty vegetables like peppers and tomatoes.

There are four main parts to go into an indoor grow light system: bulbs, ballasts, lamps, and timers. The ballast is your power source and comes in both digital and magnetic versions. Most people would recommend spending a little more on the digital version as they tend to be more efficient producing more power with less heat and run much quieter. A good simple timer isn't integral but will make your life that much simpler as you can set your lights to run for an optimum range of 15-18 hours per day without having to worry about switching your system on or off. Your plants need to sleep just like you and I and a dark period of at least 6 hours per day plays a major part in their metabolism and growth process.

There are many packaged light systems on the market today which make things easy to get up and running straight out of the box or you can buy your parts separately to create your own custom light system. If you go this route most experts recommend buying your parts together as the individual parts are made specifically for only one type of bulb preventing you from using an HID set up for Fluorescent bulbs and vice versa.

Fluorescent Lights

Fluorescent lighting produces a cool, blue light wavelength that is ideal for seedlings, young plants, and leafy green vegetables such as lettuce and most types of herbs. The most popular fluorescent light used for hydroponic purposes is the T5 which measures 5/8ths inches in diameter and usually comes in either 12" or 24" tubes. T5's are ideal for beginner hydroponic gardeners as they are the least expensive so they don't require a huge investment but more importantly don't give off much heat making them easy to work with.

While they will provide ample light for leafy vegetables they don't produce the right spectrum of light or enough lumens (light power) to support a plant's flowering/fruiting stage.

H.I.D. Lights

When you're ready to step up your plant light system or want to start growing plump, delicious tomatoes you're going to want to look at a High Intensity Discharge system. HID lights have become the standard for indoor gardening since they produce a type of light that comes very close to replicating natural sunlight but they are expensive and produce a ton of heat. There are two types of HID bulbs, each with its own strengths.

Metal Halide (MH)

MH bulbs are able to produce a light that closely mimics the power of full sunlight and is rich in the blue spectrum. This type of light is ideal for the vegetative stage of a plant's growth and produces nice, thick, stocky plants making MH a great all-purpose, go-to light.

High Pressure Sodium (HPS)

HPS bulbs are more efficient than MH bulbs allowing them to produce more lumens per watt (more light power). This allows HPS bulbs to produce a warm light in the red/yellow light spectrum which helps promote fruit formation making them ideal for the fruiting and flowering stage of plant growth.

Conversion Bulbs and Color Corrected Bulbs

Color corrected, or blue enhanced, HPS bulbs are becoming a popular option when it comes to HID lighting. These bulbs produce a light that is more balanced in the color spectrum to help promote vegetative growth as well as fruiting/flowering.

As mentioned earlier, most plant light system parts are not interchangeable so HPS bulbs typically won't work in a MH ballast and vice versa. There is a way to get the best of both worlds though and that is to use conversion bulbs - MH bulbs that work in HPS ballasts and HPS bulbs that work in MH ballasts.

LED Lights

LEDs (light emitting diodes) have been around for many years but have only recently made their way on to the indoor gardening scene. The introduction of LEDs is a very exciting prospect since they are extremely efficient, produce very little heat, and can produce very specific color wavelengths needed to promote both vegetation and flowering.

The problem, as is usually the case when adapting new technologies, is putting the proper processes into practice. Early attempts to use LEDs in indoor growing had been unsuccessful as most LED plant lights were vastly underpowered and didn't produce nearly enough lumens to promote proper vegetation, let alone flowering. Newer, more efficient models are currently hitting the market though that look very promising.

If you want to give LEDs a try make sure your selection has at least 1W per diode (e.g. a 12 diode bulb should be at least 12W).

How Much Light Size of Garden	HID Wattage	Lamp Height	*Cost Per 16hr Day
2' x 2'	175 Watt	12"	$0.14
3' x 3'	250 Watt	12" – 18"	$0.20
5' x 5'	400 Watt	18" – 24"	$0.32
7' x 7'	600 Watt	24"+	$0.48
*Calculated at $0.05 per Kwh			

T5 fluorescent plant lights are much more straightforward:

Always keep 4-6" above plants

Use 40 Watts per square foot of garden space

Use 1 4' tube per 2 sq. feet of garden space

2' x 2' = 2 4' tubes

2' x 4' = 4 4' tubes

For LED plant lights you'll want to make sure your plant lights have at least 1W/bulb. For instance a 12W bulb with 12 LED lights will have 1W/bulb. Anything less than this will not give sufficient light for your plants to thrive. A 12W bulb with 12 LED lights will be enough for a 1' x 1' garden.

Propagation

You can either start your hydroponic garden from seeds or from store bought seedlings. While seedlings provide you with a great jump start on your garden there are more disadvantages than there are advantages. First, you introduce the possibility of contaminating your garden with various pests, bugs, and diseases before its had a chance to start. Second, you'll need to thoroughly wash away all the dirt from the roots which may cause undue stress and damage to your plants as any dirt or foreign substances may clog your hydroponic set up. Most importantly, most stores only sell a limited variety of seedlings which will limit the choices you have for what to plant in your garden.

Hopefully your convinced to start with seeds so now you have to figure out how to get them to sprout! With a normal soil based garden you could just dig a little hole, plop them in, add a little water, and let Mother Nature take over. The process is pretty similar with hydroponics and there are actually a number of processes you can use.

Seed Starters

First thing is first - since you're not digging a hole in the dirt you'll need to determine what kind of home you'll give your seeds. There are several options here with the most popular being rockwool, or stonewool, starter cubes. These are literally rocks that are superheated, stretched, and spun - think cotton candy but in rock form. These make a great home for your seeds and plants as the material is inexpensive, easy to work with, chemically inert, won't break down, and works great at transporting water and oxygen.

Germination

The fact is that seeds want to sprout and, as long as they have the proper water, light, oxygen, and temperature, they will.
First you'll want to break off an appropriate number of cubes and soak them in good water for at least an hour to make sure they are properly pH balanced and to give your seeds the proper amount of water (one cube per plant per growing area). When your cubes are ready, lightly shake off any excess water and carefully place a couple of seeds in each. You'll want to place a couple in case any don't sprout and if they all do sprout you can then choose the best seedling and remove the weaker ones. If you're growing herbs you'll want to place at least 4-6 seeds in each cube and skip the thinning process since you'll want a nice, bushy garden.

Place your prepared cubes into a mini-greenhouse consisting of a domed grow tray filled with about a half inch to inch of good water or a very diluted, quarter to half strength nutrient solution. You can either use a special store bought grow tray and dome or get creative and make your own with any type of container and clear lid. Nothing fancy is required here and you can even throw each of your cubes in a sealable plastic bag if you don't have a proper container. Don't worry about light at this point since seeds are used to being buried. Just make sure you keep the temperature right around a cozy 75°F and your seeds should sprout within 2-4 days or so. If you're sprouting your seeds in colder temperatures you may want to look into either using a space heater or a heating mat to make sure your seeds get enough warmth to sprout.

Once your seeds sprout top off your tray with good water as needed - no need to add more nutrients since its just water that's evaporating. If you started with a diluted nutrient solution just top off with good water rather than continuing to add more nutrients. If you started with good water here's where you'll want to switch over to a diluted nutrient solution - just make sure its not too strong of you'll burn your little seedlings. You'll also want to introduce your new seedlings to whatever light source you plan on using whether its a nice big sunny window or special grow lights. If you are using grow lights you'll want to make sure they're not too close to your new seedlings to keep them from getting scorched by the hot lights. Then every day or two move them just a little bit closer until the lights are the proper distance.

6. Environmental and Nutritional Requirements of A Hydroponic System

Plants also need nutrition and a favorable environment to grow. Fortunately, with a Hydroponic system, we can create an almost ideal system for plants to grow to their potential. In a Hydroponic system, plants do not have to struggle to find nutrients, instead it is readily supplied to them. This helps plants utilize their energy toward growing and yielding favorable produce.

Every plant species is unique and thrives under specific environmental conditions. For example, cactus is a desert plant and grows and thrives under desert conditions. If you try to grow cactus in a cold region, it might grow but does not thrive. The same is true with nutritional requirements. These requirements vary based on the species. A tomato plant, for example, needs more nitrogen during the growing stages and less nitrogen during the fruiting stages. You can control the nutritional requirements by tweaking your nutrient solution accordingly.

ALL ABOUT NUTRIENT SOLUTIONS

As you may already know, nutrient solutions provide all the necessary nutrients for the plants to grow. You can prepare your nutrient solution at home. However, before you proceed to prepare the solution, you must understand plants' basic nutrient requirements.

Plants need two types of nutrients:

1. Micro-nutrients
2. Macro-nutrients

Nutrient/ion	Initial Normal (1)	Initial low calcium (2)	At tank change normal (3)	At tank change low calcium (4)
Nitrogen	119	127	129	129
Phosphorus	28	31	26	31
Potassium	200	233	188	231
Calcium	110	78	116	86
Magnesium	30	33	32	36
Sulfur	97	93	111	111
Sodium	72	72	86	86
Chloride	24	24	28	27
Boron	0.1	0.11	0.12	0.12
Manganese	0.04	0.07	0.04	0.05
Copper	0.08	0.08	0.07	0.08
Zinc	0.06	0.07	0.06	0.08

Micro-nutrients

Micro-nutrients are trace elements found in plants that are essential for their growth. Here is a list of micro-nutrients, their functions and problems caused because of deficiency and excess production.

Sulfur (S):

Required for: Seeding, Fruiting, protein synthesis and is a natural fungicide

Deficiency causes: Yellowish leaves and purple base

Excessive production stunts growth

Iron (Fe):

Required for: chlorophyll formation, sugar respiration to provide energy for growth

Deficiency causes: blossoms to drop from plant, yellowish color appears between the veins and leaves die at the margins

Excessive production is rare and difficult to spot

Boron (B):

Required for: formation of cell walls

Deficiency causes: poor growth and brittle stems

Excessive production may cause leaf tips to turn yellow and die

Manganese (Mn):

Required for: oxygen production during photosynthesis

Deficiency causes: Blooming fails and yellowish color appears between the leaf veins

Excessive Mn reduces Iron availability to leaves

Molybdenum (Mo):

Required for: Nitrogen metabolism and fixation

Deficiency causes: cause small yellow leaves

Excessive Mo causes leaves to turn bright yellow

Copper (Cu):

Required for: photosynthesis and respiration

Deficiency causes: leaves with yellow spots

Excessive Cu reduces the availability of Iron

Macro-nutrients:

Macro-nutrients are those that are consumed in large quantities by plants. In regular gardening, these nutrients are provided by the soil, microbes, sunlight, rain water and fertilizers; in Hydroponics, we must provide these nutrients through nutrient solution. Here are the four essential nutrients that occur naturally:

Oxygen

Nitrogen

Carbon

Hydrogen

Oxygen: required in the respiration process and helps the formation of sugar, starch and cellulose.

Nitrogen: Amino acids – the building blocks of cells – are formed using Nitrogen. Nitrogen also helps the formation of chlorophyll.

Carbon: Half the dry weight of plants consists of carbon. It is essential for the formation of chlorophyll.

Hydrogen: Plant roots absorb nutrients through a process called cation exchange. Hydrogen makes a significant contribution in this exchange and in the production of sugar and starch.

Apart from the above four nutrients, **Potassium** and **Phosphorous** also are major contributors toward plant growth. Potassium is necessary for protein synthesis and root growth. Potassium deficiency leads to fungal infection in plants. Phosphorous is a necessary element in the cells; it is present as ATP (Adenosine Tri-Phosphate). Phosphorous deficiency leads to stunted growth whereas excess phosphorous reduces the availability of copper and zinc.

NUTRIENT SOLUTION AND PH

The pH value gives us a measure of the Hydrogen ion concentration in the nutrient solution. pH value is measured on a scale of 0 to 14.

pH 7 – neutral

ph <7 – acidic

pH >7 – alkaline

A single pH value does not suffice the requirements of all the plants. Each species of plant has its pH value. A pH of <4.5 or >9.0 can adversely affect plants; in these cases, essential nutrients are locked in the solution because of extreme toxicity or alkalinity. Therefore, availability of nutrients to your plants is directly related to the pH value. An ideal pH value lies in the range of 6.0 to 7.5.

Plant names	pH range
Tomatoes	5.8 – 6.0
Lettuce	5.7 – 6.2

Eggplant	5.7 – 5.9
Peppers	5.8 – 6.2
Beans	5.8 – 6.2
Strawberries	5.8 – 6.2
Melons	5.4 – 5.6

7. Your Hydroponics Garden

Continue topping off the water in your grow tray until you can see your seedlings are a couple of inches tall and the roots start to pop out from the bottom of your starter cubes. During this time you can start preparing your nutrients reservoir so everything is nice and ready when its time to transplant your seedlings to their new home.

Growing Medium

Since you're growing with zero soil you'll need something else to hold and support your plants' roots. There are three main choices you'll have to fill your net pots including coco coir, Perlite (or silica), and LECA (Lightweight Expanded Clay Aggregate; also known as Hydroton). All three have the same basic characteristics in that they are all made of chemically inert materials. This is important for your growing medium of choice as these substances won't react with or contaminate your nutrient solution or affect your plants in any way.

Coco Coir is made from the outer brown husk of coconut shells. The substance is basically a bunch of tiny sponges that allow it to hold many times its weight in water making it a great growing medium for your hydroponics garden. Its usually sold in brick form and has to be reconstituted in water before using. While coco coir is great for aeration and water retention it does break down easily over time and is best used in conjunction with another growing medium in a 50/50 mix.

Perlite, or silica, is basically popcorn made with flakes of glass and is made by superheating specs of glass (silica instead of corn) until they expand (like popcorn). It is very light and inexpensive, holds water well yet still provides excellent drainage. These characteristics make silica a great standalone growing medium with the only drawback is that it can be a little too light and is easily washed away in some systems and is most often used in mixtures as additional filler.

LECA, more commonly know by Hydroton, is made in a similar way as perlite - think popcorn made from little clay balls. These clay balls are very light, do a good job of retaining moisture, provide great aeration, and won't wash away in your reservoir. Those traits, along with the fact that they can be cleaned and reused over and over make LECA a top choice for most hydroponic gardens. If you do choose to go with clay just make sure to thoroughly rinse before your first use to wash any loose dust from them as they come out of the bag pretty dirty.

Aeration

The final component of your hydroponics garden is going to be an air pump and air stone (with some connective rubber tubing). Proper aeration of your nutrient solution serves two main purposes: 1) inhibit growth of mold, algae, and bacteria; 2) deliver oxygen to your plants' roots to help promote growth.

Prepping Your Reservoir

Once you've decided on what nutrients and growing medium you're going to use, have your lighting and aeration in place, and have access to good water its time to put it all together!
Start by mixing enough of your nutrient solution to fill your reservoir so that it reaches about 1/3 to 1/2 up from the bottom of your net pots. Remember to always use pH balanced good water and to follow any nutrient mixing instructions exactly. Connect your air pump and air stone using an appropriate amount of rubber tubing and place the air stone in the center of your reservoir.

Transplanting Your Seedlings

Depending on the size of your net pots and how nice you'll want to make things look - fill your net pots with enough of your growing medium so that the top of your starter cubes will be aligned with the top of your net pot. Carefully place your starter cube and seedling in the net pot and continue to fill the net pot with your growing medium to create a snug fit. Place your net pots in your slotted reservoir lid and voila - you've just completed your hydroponics garden and are on your way to growing fresh produce in the comfort of your own home!

Ongoing Maintenance

Over time the water in your nutrient solution will evaporate and you'll need to top off to bring your water level back up. Make sure to top off with good water only and never add more nutrient. The water loss is purely due to evaporation so your nutrient ratios will remain the same. Adding more nutrients will only increase the concentration and could lead to nutrient burn or other harmful issues.

There are different rules of thumb as to when and how often you should change your nutrient solution. Some of the more ambitious hydroponics gardeners mix a whole new solution ever week while others have great results changing their solution every 2 weeks or even once a month. If you're looking for a true hands off approach you can even use a 50% rule and only mix a new nutrient solution once you've topped off 50%

of your entire reservoir with good water. Experiment with different schedules and pick the one that best suits you. Just remember that your hydroponics garden is supposed to be fun - not a chore!

8.   Overview Of The Different Hydroponic Systems

There are six basic types of hydroponic growing systems, each with their own unique benefits that can be adjusted to suit the grower's needs. As such, they can also be combined and customized to fit into different lifestyles. Overview of the Different Hydroponic Systems

Drip System

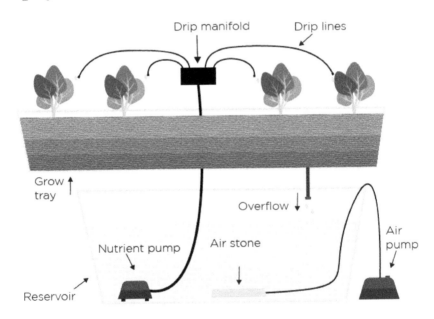

Probably one of the most popular of all the hydroponic systems is the drip system. It is used in commercial applications, on city rooftops, and even in indoor apartment gardens. It is by far the most versatile and efficient system that is also quite easy to maintain.

Benefits of the Drip Hydroponic System

- The drip system is relatively easy to build and maintain.

- The drip system is a good choice for plants with larger root systems.

- The drip system offers the ability to grow plants all year round.

- The drip system has a more versatile watering and feeding schedule.

- The drip system has a very water-efficient irrigation system.

- The drip system is relatively cheap to maintain.

How a Drip Hydroponic System Works

The drip system works by delivering water-based nutrients to the plant's root system by a low-flow method called drip irrigation. It is a very water-efficient system that has little-to-no waste that would normally occur due to evaporation. It provides a controlled and steady slow drip of moisture to the base of the plants instead of showering them from above.

The Working Parts of the Drip Hydroponic System:

Grow Tray

The grow tray is a shallow tray that sits on a shelf or a stand that is around 6 inches above the reservoir. This tray houses the grow pots that contain the plants. The drip manifold is fed through from the reservoir to the grow tray.

Drip Manifold, drip hoses, dripper stick, and drain filter

The drip manifold is usually a specially bought manifold that sits in the grow tray. The manifold has drip emitters that feed the drip tubing. The drip tubing has a dripper stick attached to the end which gets inserted into the growing medium in the pot next to the plant's roots. The dripper stick is what delivers the water-nutrient solution to the roots.

The drain filter is what filters out the overflow and unused solution as it attaches to an outlet pipe that feeds back into the reservoir.

Pots

These are the individual pots that will hold the growing medium and the intended plant. They will stand on the grow tray. They should be of adequate size to house the plants and growing medium. As plants stand in individual pots, it allows for a mixed growing environment where you can grow different types of plants housed in one hydroponic system. In a standard system, the plants would have to have similar watering, pH and nutrient requirements. Commercial growers with special pumps and emitter have a more flexible growing environment.

Growing Medium

The growing medium is an important part of the drip system. It has to be able to provide excellent grounding support for your plant. Most importantly it has to be able to absorb and hold enough water, nutrients, and oxygen to sustain the plants while supplying adequate drainage. The growing medium is inserted into the pots that house the plants in place of potting soil

Growing Mediums that Work Well with the Drip System are:

- Coco coir

- Lightweight expanded clay aggregate

- Perlite-vermiculite mix

- Rock wool

Reservoir

The reservoir, which holds the nutrient-rich water that will feed the plants in the grow tray, is situated directly below the table or the stand that houses the grow tray. It is usually a lot deeper than the grow tray and will house the nutrient pump (with a timer), overflow hoses, and usually an air stone attached to an outside air pump to aerate the water.

The reservoir houses:

- The nutrient pump runs the drip manifold, which protrudes up and through the growing medium. Individual drip lines usually extend from the manifold like a sprinkler system with long sleeve arms. These drip lines each serve an individual plant. Most of them come with a regulator that can be individually adjusted for more feeding control. This makes it easier to grow different types of plants in one grow tray.

- The overflow or return pipes are what drain unused or excess water and nutrients back into the reservoir to be recycled or flushed.

- The air-stone and its air-pump are needed to oxygenate the water by creating little bubbles of oxygen in the water. The pump creates the bubbles and the stone is used to disperse the oxygen into the tank. The size of both the stone and pump needed for the system is dependent on the size of the reservoir tank.

The Two Types of Drip-System

- Non-recovery drip system

In this system, nutrients and water are not recycled as the system has very little to no wasted solution. This is due to the use of sophisticated cycle timers set to a very specific and precise watering schedule. This schedule is defined down to a precise second if need be. It moistens the growing medium around the plants just enough to afford them the nutrients they need. This means there is not a lot of, if any, wasted solution.

As the water nutrient solution is only used once, the system is not as maintenance intensive as the recovery drip systems. Since the solution in the reservoir is not being recycled, the pH and nutrient levels remain constant. Although, it is good practice to regularly check these levels and the tank will have to be refilled when necessary.

It is also advisable that the growing medium for the plants is completely rinsed out with clean fresh water from time to time. This is to stop a nutrient build-up that could jeopardize the system's precise nutrient levels that need to be maintained.

This system is very popular and widely used by commercial growers because of its versatility and ability to control watering down to the second. As it relies on technology and some specialized equipment it can be quite costly, making it more suited to commercial growers.

- Recovery drip system

This is the most efficient system for people who want to have garden hydroponic drip systems. In a recovery system, the water and nutrients are reused over and over again. The water-nutrient solution is slowly dripped onto the base of the plants where it will slowly trickle through the growing medium. This allows the roots to take what they need from the solution. If there is any unused solution leftover, it will trickle back down to the reservoir to be reused.

This system is also known as the recycling system, and while it may be cost-effective saving on nutrients and water, it requires quite a bit of maintenance. Each watering cycle, the plants will take the nutrients they require from the solution which will change both the nutrient level and pH balance of the solution. As such, it needs to be constantly monitored and adjusted when needed.

The reservoir will need periodic emptying out and refreshing of the water-nutrient solution. This is usually the time where growers check their mini pumps, replace air-stones, and any tubing that may require renewing.

Plants to Grow in a Drip System:

The drip system is quite a versatile one and can be designed to grow a number of plant varieties. As their drip can be controlled per plant, they can get as much or little water as the plant requires to thrive. Thus, the drip system caters to both thirstier plants and those that like a drier climate. This system can also be used for larger plants and trees.

These include:

- Most herbs

- Most leafy greens like lettuce, spinach, chard, etc.

- Cabbage and broccoli

- Most commercial vegetables

- Flowers

- Even some fruits like tomatoes, peppers, strawberries, and bananas.

Ebb & Flow

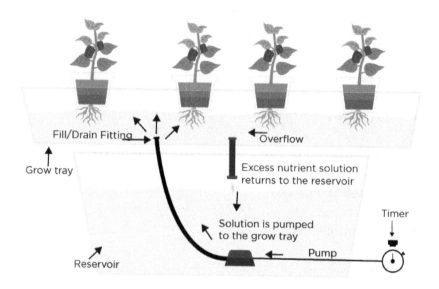

The Ebb and Flow system is probably the system most people associate with hydroponic growing. It is an extremely versatile and cost-effective system to get started but it is not the best for beginners. It is more an intermediate level system.

Benefits of the Ebb & Flow Hydroponic System

- The ebb and flow system is cost-effective to set up and maintain.

- The ebb and flow system does not require expert knowledge to grow healthy plants. Although it will take some training to start.

- The ebb and flow system provides healthy plants all year round if properly set up and maintained.

- The ebb and flow system have a simple watering solution that provides the plants with just enough nutrients.

- The ebb and flow system very water-efficient irrigation system that can reuse water-nutrient solutions.

- The ebb and flow system offers the ability to easily change plants in the planting tray without affecting the rest of the crops.

How the Ebb & Flow Hydroponic System Works

The ebb and flow system may not be the easiest of systems to set up, but it is not the hardest either. Once it is set up it is really easy to maintain and with a bit more experience, a person will find they start to save on various materials. This is especially true if you build the system yourself thus enabling you to scale the project to whatever size you wish.

As the name implies the ebb and flow system means that the plants are periodically flooded with a water-nutrient solution which is then drained away.

It is also called the flood and drain system and consists of two phases:

- Flow (flood)

The system works on a timer that will start the watering cycle. This is done by means of a pump that pushes the water-nutrient solution up into the growing tray. This floods the roots until it reaches the water limit.

The nutrient-based solution is circulated around the tray for a set time.

- Ebb (drain)

Once the roots of the plants have been flooded with the water-based nutrient solution for the set time the timer will go off again. At this point, the pump will stop, and the solution will drain back into the reservoir through the outlet/overflow pipe. This allows for the water-based nutrient solution to be re-used a few times, provided the nutrients and PH levels are kept at the optimum levels.

The Working Parts of the Ebb & Flow Hydroponic System:

Grow Tray or Flood Tray

The grow or flood tray of an ebb and flow system is a large tray that is not too deep. The tray usually sits a couple of inches above the reservoir, on a stand, or on a table. This tray has inlet and outlet openings to accommodate the pump and drainage filter.

Water-Nutrient Inlet/Outlet Pipe, and Drain Filter

The reservoir feeds the flood tray with an inlet pipe that is fed from the submersible pump.

A drain filter is used to drain the water back into the reservoir.

Pots

These are the individual pots that will hold the growing medium and the intended plants. They need to be twice as deep at the flood tray. The base must be large enough to accommodate the growing medium. There must be adequate openings at the bottom for the nutrient solution to flow up through to reach the roots.

Growing Medium

The growing medium for the ebb and flow system has to be strong enough to support the plants. It must not retain too much moisture and has to drain well enough to allow the roots to dry out between flow cycles.

Growing Mediums that Work Well with the Ebb and Flow System are:

- Clay grow stones

- Rinsed gravel

- Sand

- Rock wool

Reservoir

The reservoir is usually twice as deep as the grow or flood tray. It is what holds the water-nutrient solution. This is the solution that gets pumped up into the flood tray and through to the roots of the plants. The reservoir connects through to the flood tray by the fill pipe and draining tubes.

The reservoir houses:

- The submersible nutrient pump that has a timer that gets set to flood the grow tray and turn off to allow the water to drain off.

- The overflow or return pipe is what allows for the ebb of the nutrients back into the reservoir.

- The air-stone and its air-pump are an optional extra. They give the water a bit more oxygenation but as the system is already pumping water up and around it should have adequate oxygenation.

Plants to Grow in Ebb and Flow System:

Plants that do not mind a lot of water and grow quite fast are best suited to the ebb and flow system.

These include:

- Basil
- Beans
- Beets
- Broccoli
- Cabbage
- Chard
- Chives
- Cucumber
- Kale
- Lettuce
- Mint
- Watercress

Nutrient Film Techniques (NFT)

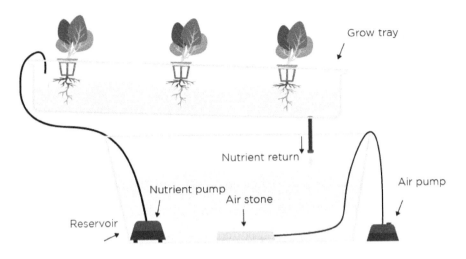

The nutrient film technique (NFT) is a very popular growing technique amongst hydroponic hobbyists and some commercial growers as it is a really versatile system. It has a lot of similarities to the ebb and flow system. But where the ebb and flow floods and then drains, NFT systems run a constant flow of nutrients over the roots of the plant.

Benefits of the Nutrient Film Technique Hydroponic System

- The NFT system a low consumption of nutrients and water.

- The NFT system's constant water flow stops the salt build-up at the roots, it is easy to clean, and has a lower risk of contamination.

- The NFT system needs little to no growing medium.

- The constant flow of water and nutrients creates a thin nutrient film that keeps roots healthy.

- The NFT system is easy to expand because it is a modular system.

- The NFT system is quite easy and not too expensive to main.

How the Nutrient Film Technique Hydroponic System Works

NFT system has a reservoir that contains a water-nutrient solution. This solution is pushed up through a pump into the growing tubes. The growing tubes are usually suspended at an angle so the water flows down towards the outlet pipe that feeds back into the reservoir. The nutrient water solution is reused over again until it is time to change it. A continuous thin film of nutritious solution constantly flows over the roots nourishing them.

Most growers use flat bottom tubing systems that have grooves that run lengthwise along the growing tubes. This helps stop the system from damming up as the water runs beneath the roots.

There should be around a 1:30 ratio which means that there is a one-inch slope for every 30 inches of horizontal tubing length. Keeping the grow tubes flexible means they are easily adjustable as the plant's root systems grow, ensuring that the gully's do not get clogged by the roots.

Shorter tray runs are best to ensure all the plants receive an adequate nutrient solution. Because the water flows down, the plants at the top of the run tend to get most of the nutrient content. A run that is too long may cause the plants at the lower end of the run to become nutrient deficient. That is not to say that long runs cannot be as successful, just that they require a lot more attention to the PH and nutrient balance.

The Working Parts of the Nutrient Film Technique Hydroponic System

Grow Tubes or Channels

NFT systems use channels or grow tubes where most other hydroponic systems use trays or buckets. These tubes make it easier to position them at an angle in order to get the correct water-nutrient solution flow over the plant's roots.

Although some DIY growers use round PVC piping, it can cause uneven watering. A channel with a flat bottom, preferably with shallow, lengthwise grooves at the bottom, is a much better option.

Water-Nutrient Feeding Pipe, and Drain Filter

The reservoir feeds the water-nutrient solution into the grow channel by a pipe attached to a pump. This pipe feeds the water into the channel at the raised end, which makes it flow downwards, constantly feeding the roots as it heads to the outlet drain. The drain filter on the lower end of the channel allows the water to run back into the reservoir to be recycled.

Pots

Most NFT systems put the plants directly into the channel openings, in caps that can gently hold the plant in place. For plants that need more support, seedlings are planted into net pots and they are put into the channel openings. The roots of the plant need to be free to dangle and grow in the channel. This does, however, mean that root maintenance is required in order to stop them from becoming tangled and clogging the system.

Growing Medium

A growing medium is not really required with the NFT system unless it is for seedlings. If a growing medium is to be used it should not be a lot and the roots should be able to protrude through it enough to reach the film.

Growing Mediums that Work Well with the Nutrient Film Technique System are:

- Lightweight expanded clay aggregate

- Diatomite

- Gravel

Reservoir

The reservoir is usually situated beneath the grow channels and the size of it depends on the number of channels it feeds. It is connected to the growing channel(s) by the feeding pipe on the high end and the drainage pipe at the lower end.

The reservoir houses:

- The submersible water pump that runs the water-nutrient solution feeding pipe. This pump runs steadily all the time producing a constant stream of water and as such, is not attached to a timer. This can make the system vulnerable as it is completely reliant on the pump running properly. Blockages or downtime due to a systems failure or power outage could cause problems for the system. The pump should be checked on a regular basis and a backup system installed in case of emergencies.

- The drainage pipe feeds down into the reservoir from the lower end of the channel. This pipe makes sure that water constantly flows through the system and is recycled.

- An air-stone along with a pump to generate oxygen is a good idea and adds much-needed oxygen to the water. The size of the stone depends on the size of the hydroponic system and sometimes two may be needed. The shape of the stone is a purely personal preference although some say different shapes tend to work better than others.

Plants to Grow in a Nutrient Film Technique System:

NFT systems are best suited to the growing of plants that are leafy like lettuce and that have a shorter growth time.

These include:

- Basil

- Broccoli Rabe

- Chard

- Chives

- Dill

- Lettuce

- Spinach

Water Culture

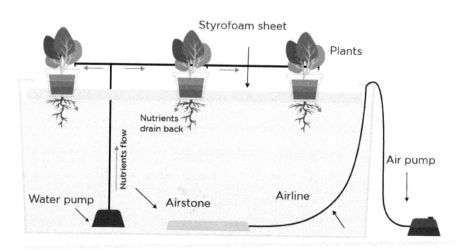

The water culture system is one of the easiest of the six systems to both build, learn, and maintain. It is used with home growers and commercial growers, as the system can be scaled to any size. Plus, there are a lot of innovative ways to design various growing environments.

Benefits of the Water Culture Hydroponic System

- The water culture system is very easy to design and set up.

- The water culture system is not too expensive to get started with.

- The water culture system is quite easy to maintain, and upscaling is not too expensive.

- They are good for just about any size plant.

- They are great for beginners to start with.

- Aerated roots of the system make for faster growth and healthier plants.

How the Water Culture Hydroponic System Works

The water culture or deep water culture is a simple hydroponic solution whereby the plants sit in a bit of growing media in a netted pot. This pot is held in position by a lid that is placed on top of a water-nutrient filled reservoir. The roots of the plant hang freely in the nutrient solution. The water is kept aerated by a pump and air stone.

The Working Parts of the Water Culture Hydroponic System

Reservoir Lid

The reservoir lid is usually made from plastic or polystyrene. It has cup hole openings in it. The size and amount of opening vary upon the needs of the growing environment. The lid must fit comfortably on top of the reservoir and not be too thick or too thin for that matter. The openings on the top of the lid must be just big enough to be able to suspend the plant baskets. The size will vary on the growing requirements.

Pots

The plants will need to be placed into netted pots that allow for their roots to hang freely. The size of the pots will depend on the type of plants being grown.

Growing Medium

The best growing medium for water culture has low-water retention. This is mainly used for starter plants and the top half of the medium needs to stay dry.

Growing Mediums that Work Well with the Water Culture System are:

- Grow rock

- Rock wool

Reservoir

The reservoir is the tank that the grow lid will cover. The tank is filled with water and the required nutrients for the type of plants growing in it. The water should be filled to a level where the roots of the plant sit comfortably in the solution. The water to nutrient ratio depends on the nutrients being used, the size of the growing environment and the plants being grown.

- As the plants are suspended directly into the water solution, a water pump or drainage filter is not required.

- The system does need an air stone and an air stone pump. The pump creates the tiny oxygen bubbles which are dispersed into the tank by the oxygen stone. The size of the stone and pump once again depends on the size of the growing environment.

- The reservoir does require some maintenance and the water-nutrient solution needs to be completely cleaned out and refreshed every other week or so. The water temperature, nutrient and PH balance of the water also needs to be closely monitored in this system.

Plants to Grow in a Water Culture System:

Deep water culture plants cannot be too top-heavy as they usually need to be supported in some way. They must also be plants that do well in a wet environment, i.e. thirsty plants. Plants like Rosemary that do not like too much water and prefer their environment to be a bit drier will not do well in the water culture environment.

Thirsty plants include:

- Basil

- Broccoli

- Cabbage

- Chard

- Kale

- Lettuce

- Okra

- Sorrel

Aeroponics

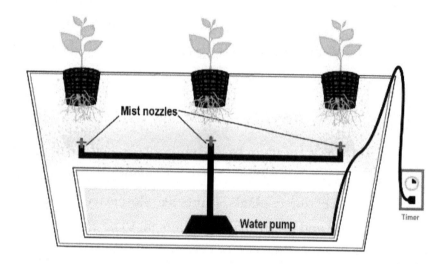

Aeroponics is one of the most efficient of the hydroponic techniques as it uses a misted spray technique to feed nutrients directly to the plant's roots. This is accomplished without the use of a growing medium and affords the roots maximum oxygen exposure.

Benefits of the Aeroponics System

- Aeroponic systems afford the roots a higher oxygen rate which encourages rapid growth and healthier plants.

- They are a good choice for limited space.

- They use less water and nutrient solution.

- They need very little, if any at all, growing media.

- They are a lot easier to harvest than most other hydroponic crops.

- They are relatively easy to maintain.

How the Aeroponics System Works

Out of all hydroponic systems, aeroponics is the most difficult to set up, even though the actual concept is a relatively simple one. The plant is suspended with the roots exposed to the water-nutrient solution in the form of a misting spray. As the roots are suspended, they are able to absorb much more oxygen. Due to the quality of the water-nutrient content and high oxygen exposure, the plants tend to grow a lot faster and are a lot healthier.

The Working Parts of the Aeroponics System

Growing Chamber

This is an enclosed chamber where the plant's roots are hung. It has to be airy enough to let in enough oxygen but at the same time keep pests out. It must also be able to keep in the humidity, water, and nutrients.

Water-Nutrient Misting Piping and Spray Nozzle(s)

The pump will send water through tubing/piping to the misting nozzles for sprinkler heads. These sprinklers spray a delicate stream of the water-nutrient solution onto the plant's roots.

Pots

The aeroponics system does not use pots as the plants are suspended by growing chambers.

Growing Medium

Aeroponic systems only require a growing medium for starter plants such as seedlings and cuttings. Plants that are being transplanted may also require a growing medium. Once the plant starts to mature or liven up and its roots start to form it will no longer require a growing medium. For the plants that do need one it should be a medium that does not absorb or hold onto the water or the plants may become susceptible to stem rot.

Mature plants in an aeroponic system do not require a growing medium.

Growing Mediums that Work Well with the Aeroponics System are:

- Grow rock

- Rock wool

Reservoir

This is a container or small tank that holds the water-nutrient solution. The size and type depend on the growing environment and what type of aeroponics system being used.

There are two types of aeroponic systems:

- Low-pressure system

This is the most common of the two and is used by home growers or hobbyists. It uses a normal pump which operates low-pressure misters that have a light spray and not a "true mist". This system is the least expensive, and easier to set up and maintain than the second option. It delivers larger drops of water than high-pressure systems as well.

- The reservoir for the low-pressure system contains the water-nutrient solution and has a standard submersible pump attached to a cycle timer.

- Tubing attached to the pump runs up to the root zone chamber for the sprinklers.

- High-pressure or True mist

This is a system were an actual mist floats in the air. It is a more effective way of delivering moisture and nutrients to the root system. This is a fine spray with small droplets of water.

- A high-pressure water pump is capable of delivering the necessary misting spray.

- The tank should be able to hold around 60 psi and the system will need specialized misters that can spray only a breath of fine moisture.

Plants to Grow in an Aeroponic System:

Aeroponic gardening system is one of the most versatile and can grow a wide variety of plants.

These include:

- Basil

- Chives

- Grapes

- Kale

- Lettuce

- Mint

- Oregano

- Rosemary

- Sage

- Tomatoes

Wick Irrigation

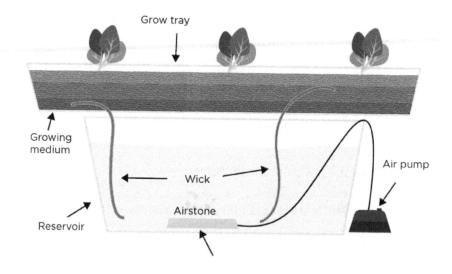

The wick irrigation system is probably the most basic and simplest of all hydroponic systems. That is why they are the best systems to teach beginners and kids that want to learn about hydroponic growing. The wick system is what is known as a passive system as it does not have any pumps (unless if you have an air stone) or drains, etc. This also makes it one of the most dependable of the systems as it is not reliant on any technology or electricity. All it needs is a bit of maintenance and regular check-ups to have you growing plants in no time.

It is also ideal for people who do not have a lot of space for a garden, as the system is very versatile. It is not too expensive to maintain and is extremely easy to get up and running.

Benefits of a Wick Irrigation Hydroponic System

- The wick irrigation system is the easiest of the systems to get started with.

- The materials are easy to source and not too expensive.

- They are very versatile in that they can be a small one with a single plant to a much larger commercial type of system.

- They are ideal for an apartment hydroponic garden.

- They are the best systems for beginners to learn with.

How a Wick Irrigation Hydroponic System Works

The wick irrigation system is best suited for faster-growing plants like lettuce. Although they are easy and cheap, they are not great for growing larger plants that need a lot of water or have finicky watering or nutrient needs. The wick irrigation system works by having a wick or two made from certain absorbent materials. The wicks have one end tucked into the plants growing medium close to the root. The other end of the wick gets suspended in the water-nutrient solution found in the reservoir. The water is fed to the plant through a capillary action that keeps the roots moist in nutrient-rich water solution.

The Working Parts of the Wick Irrigation Hydroponic System

Grow Tray

In a wick irrigation system, the grow tray has the growing medium spread in it with the intended plants planted directly into the medium.

Wick

The wick can be a length of felt, some string, or a rope made of cotton, as long as it allows the nutrient solution to be transported via capillary action to the roots. The number of wicks and length of the wicks depends upon the size of the system.

Pots

Most wick systems do not use pots or netted pots as the plants get planted directly into the growing medium in the growing trays.

Growing Medium

The wick irrigation system requires a medium that can absorb and maintain moisture in order to ensure the plants are suitably fed for optimum health.

Growing Mediums that Work Well with the Wick System are:

- Coco coir

- Perlite-vermiculite mix

- Rock wool

Reservoir

The reservoir needs to be able to efficiently supply the number of plants in the growing arena. It must also be big enough that it does not need to be refilled every other day. The reservoir does have to be cleaned out every now and then to keep the system fresh and working properly.

- The wick system does not need a pump as it is a passive system.

- There is no feeding pipe, so the reservoir is attached to the growing tray by the wicks.

- An air-stone would probably be a good idea in a wick system. As there is no active water flow movement, the stone and its pump help with oxygenating the water to ensure your plants are receiving enough oxygen.

Plants to Grow in a Wick System:

The best plants for a wick system are fast-growing plants that are not very thirsty.

These include:

- Basil

- Lettuce

- Rosemary

Media Beds

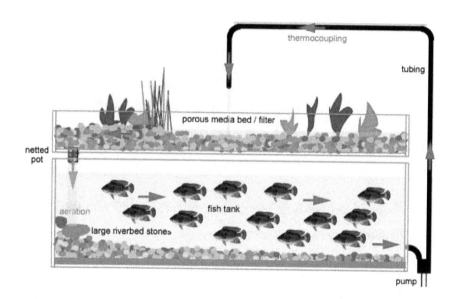

Media bed growing is a form of aquaponics that utilizes the symbiotic relationship between fish and plants to water and nourishes soilless crops. The result is fresh healthy food that is grown in an environmentally friendly way.

Benefits of a Media Bed Aquaponic System

- The media bed growing system produces wholesome organic food.

- The media bed growing system cuts down on waste since it's a recycling type system.

- It is a great system for maintaining a fish farm. If not, the fish make a great center point for the hobbyist.

- It provides a cleaner and healthier growing environment with no soil.

- Media bed aquaponics is not that hard to set up.

- Smaller gardens can be set up in apartments or rooftop gardens.

How a Media Bed Aquaponic System Works

Media bed aquaponics works on the same principle of the ebb and flow hydroponic system. Only the plants are being flooded from the water of a fish tank. The water from the fish tank contains all the bio-nutrients the plants need from fish waste products. The plants, in turn, break down the waste in the water cleaning it to be returned to the fish tank.

The Working Parts of the Media Bed Aquaponic System

Media Bed

The aquaponic system uses a media bed in which the entire base is filled with a growing medium. The plants are planted directly into the bed and the system is watered by a pump system that delivers nutritious water from a fish tank. The bed will have the water hose at the one end and a bell siphon at the other end. The bell siphon is what delivers the water back to the fish tank. The water that is drained back has had any waste product broken down in the plant bed. There are times when some growers may introduce specific worms into the gravel of the plant bed to aid in the waste break down.

The size and amount of media beds that can be serviced by a fish tank are dependent upon the size of the fish tank.

Watering Pipe and Outlet Flow Pipe

The watering tube must run from the water pump in the fish tank onto the top or the media bed. The outlet flow pipe will run from the bell siphon back into the fish tank. This has to be set in a way that allows for gravity to drain the water back into the fish tank.

Pots

The plants are planted directly into the grow media in the beds and do not need pots.

Growing Medium

The growing medium needs to be porous and able to maintain enough moisture between watering as well as firmly hold the plant's roots. The growing medium has to completely cover the growing bed.

Growing Mediums that Work Well with the Media Bed Aquaponic System are:

- Pea gravel

- Expanded clay pebbles

- Perlite

- Peat moss

- Coco coir

- River rocks

Fish Tank

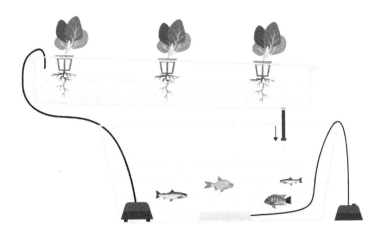

Unlike hydroponics, aquaponics is fed from the nutritious water of a fish tank. The fish tank should be set up with the usual gravel stones, air stone, pump, and of course the fish. It will also need the water pump that feeds the media bed(s) and must be powerful enough to ensure that the water is efficiently reaching and flooding the bed(s).

The drainage system will depend on how the beds are set up, but drainage usually goes through a bell siphon and uses gravity to pull the water back into the fish tank.

The size of the fish tank and the fish that are in the tank are purely a personal choice. A rule of thumb for the size of the tank to media bed ratio is:

- Every 10 gallons of fish tank water can support up to 2 square feet of growing media.

How to Choose a Fish Tank for Aquaponics

When choosing a fish tank for aquaponics you do not necessarily need a big fancy tank. There are many options that are just as good as the top of the line ones. You can even make your own tank from a rainwater tank, etc.

When choosing a fish tank for an aquaponics system, consider these points:

- The tank must be watertight.

- The tank's plumbing fittings must be secure with no leaks or misaligned fittings.

- The tank must not have any toxic material and needs to be sterile.

- The tank must not be made of metal as metal is prone to rust. Corrosion has an effect on the water's ammonia and pH balances.

- The tank must have no plants or other items that can affect the water's pH balance.

- The tank size must be able to comfortably house the type of fish that are going to be used for the aquaponics system. Check how big each species of fish gets before committing to a fish tank.

- Any shape of the tank will do, this is more a personal choice and does not in any way affect the aquaponics system.

- Choose a tank that will fit neatly into the allocated space where the aquaponics system is going to go.

- A pond can be used but you will have to cover it with a pond liner or pond skin. These are usually puncture resistant and help to keep the pond from getting infected by various bacteria, algae and other substances that can be detrimental to an aquaponics system.

Choosing the Best Location for the Aquaponics System

- Only areas, like Florida, that have year round good weather can accommodate an outdoor aquaponics system. You cannot grow vegetables or have fish outside in very cold temperatures.

- If you live in an area that has great summers and cold winters, you will still have to look at housing an aquaponics system in an indoor environment.

- It is possible, to have a system that is outdoors during the warmer months and indoors during the colder ones. There are ways around this; for instance, the system can be moved inside if it is portable or a temporary structure could be set up around the system. However, this is not very cost effective.

- The best solution for an aquaponics system is to have in a greenhouse type environment.

- If you need to have the aquaponics system inside, like in an apartment building, it will have to be positioned where it can benefit from the best lighting. If there is no natural light, artificial lighting will have to be used. If it is possible, combining as much natural light as possible with artificial light is the most beneficial.

- Although it is easier to work with the fish tank does not need to be right on top of or directly next to the grow beds. However, it must be positioned on something that is stable and will not collapse.

- The position of the tank is also dependent on the type of aquaponics system chosen, as deep water culture, media base and nutrient film technique all have different setup requirements.

What Needs to be Included in the Fish Tank:

The most obvious things an aquaponic system needs is some water and fish! But there are few more things that need to go into the tank. These ensure that the system works and that the fish tank produces the nutrients that are required to grow the plants and keep the fish healthy.

- An Air Stone is needed to help with water filtration which is crucial to the success of an aquaponics system. An Air Stone circulates both oxygen and nutrition around the fish tank, which is vital for the water that flows onto the roots. An Air Stone can also ensure that the nutrient solution in the water lasts longer.

- A Pond Filter helps to keep the fish tank cleaner for longer, and stops unwanted bacteria build up. There are two types of pond filters that should be considered:

  - The Biological Filter helps to maintain the water quality in order to keep the fish healthy.

  - The Ultra Violet Clarifier is the filter that keeps away the agents that make the water green.

- The best filter to get is one that does both biological filtration and has an ultraviolet clarifier. A lot of the filters available on the market today do both but it is always prudent to check before buying one.

- A Pond Monitor is a wise investment to make as it continuously monitors the fish tank for leaks, changes in pH and ammonia levels, and water temperature. Some pond filters come with built-in pond monitors and are worth buying. Having a monitor can prevent the sudden unexplained death of the fish in the tank or of potentially losing an entire plant crop.

- A good Water Pump is needed to ensure that the system has enough power to be able to release the water to the crop. It must be watertight, and all the parts must fit snugly together. The pump has to be compatible with aquaponics systems as there should be no toxic components on it. The best pumps have no copper parts that are exposed to the water, are energy efficient, and can work on a timer.

Taking Care of the Fish Tank

- In an aquaponics system, the water is recycled over and over again. The waste matter produced from the fish is used to feed the garden; the garden, in turn, cleans the waste from the water and sends clean water back to the fish tanks. Although the fish are getting cleaned water, the water does start to lose nutrients and still starts to become unhealthy for the fish.

- The pH and ammonia values in the water have to be checked on a regular basis and the fish tank has to be kept clean. There are some tanks snails that can help to clean away bacteria but before putting any such mollusk in the tank, check to ensure they will not affect the rest of the aquaponics system.

- The food the fish are fed can also make a big difference to the water's pH levels and the fish should only be fed food that is recommended for the aquaponics system.

- If the bottom of the tank is lined with gravel, this must be kept clean and changed often. Fish food and some fish excretion can start to build-up on the fine stones in the system.

- Check the fish to make sure they are all healthy as one unhealthy fish can lead to problems for the rest of them. Remove the fish immediately, and clean out the fish tank.

- Remove any dead fish immediately and clean out the fish tank. Replace all the gravel and any air stones.

Fish that Do Well in Aquaponics:

Just like some plants do well in various hydroponic, aeroponic, and aquaponic environments, there are certain fish species that tend to be better suited for the process.

Freshwater fish that like warmer water do better with crops of leafy vegetables such as herbs, lettuce, and cabbage. Tomatoes do better in larger systems that have a lot more fish in the tank, etc.

Some fish species to consider for aquaponics (depending on the size of your growing area and fish tanks):

- Angelfish

- Barramundi

- Blue Gill

- Carp

- Catfish

- Crappie

- Goldfish

- Guppies

- Koi

- Largemouth Bass

- Pacu

Plants to Grow in Media Beds:

Plants that do well with most aquaponic systems:

- Arugula

- Basil

- Chard

- Chives

- Kale

- Lettuce

- Mint

- Watercress

Plants that need a fish tank that has a lot of fish in it:

- Bananas

- Beans

- Beets

- Broccoli

- Cabbage

- Carrots

- Cauliflower

- Cucumbers

- Microgreens

- Onions

- Peppers

- Radish

- Squash

- Sweet Corn

- Tomatoes

Vertical Gardens

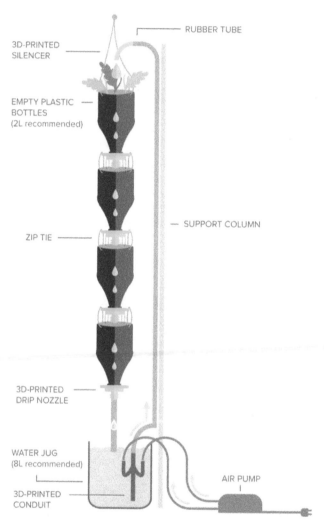

Humans invented high rise buildings in order to maximize space for both office and home solutions. Instead of spreading out they found the only option was to go up. This is also true for growing space. When you do not have a lot of gardening room to spread your plants out the only other logical way to go is to stack them up.

Places with poor soil and limited space have an option to not only stack up but also to grow those stacked up plants hydroponically. Vertical farming mixed with hydroponics is what brought about the concept of vertical hydroponics. It is also known as vertical gardening, tower hydroponics, or tower garden systems. This allows for stacked level growing where plants are grown at each level. These stacks can be several layers high.

If this concept sounds vaguely familiar to you it is because it is not a new one and has in fact been around since ancient times. One of the many ancient wonders of the world was the Hanging Gardens of Babylon from around 500 BC. It consisted of trees, shrubs, and flowers hanging at different levels in this huge garden.

Benefits of a Vertical Hydroponic Garden

- Soil systems are heavy and weigh down vertical garden. However, there are hydroponic systems that are lightweight and can reduce the garden weight by around 28 to 30 percent.

- They save a lot of space and allow for more plants to be grown which is ideal for the urban grower.

- Hydroponics offers a faster growth rate and you can plant more plants for a greater yield in a smaller space.

- Little to no water and nutrient waste occurs due to the recyclable nature of the system.

- They are aesthetically pleasing, can be soothing and offer other health benefits besides nutrition.

- They are easier to maintain than traditional soil-based gardens.

How a Vertical Hydroponic Garden Works

There are a few hydroponic techniques that could be used for a vertical garden. The two most obvious being the aeroponic systems and the NFT system, although the closed system with a constant flow feeding the roots is probably the better option of the two.

In a vertical tower, a grow tube is used to hold the plants, while a pump pumps the water-nutrient solution out through a single tube that is positioned at the top of the grow tube. The water then runs down through drainage holes positioned over the roots of the plants. This feeds the plant's roots as gravity pulls the water down with the excess draining back into the reservoir.

The Working Parts of the Vertical Hydroponic Garden

Grow Tube

The grow tube is usually a long PVC pipe or similar. The size is dependent on the type of plants being grown and how many layers of plants there are. The grow tube will have an opening up the sides for the net pots.

Inside the grow tube, the net pots must not touch each other, with a blind stop on top of each pot location to stop the nutrients from splashing out. At the top of the tower will be the sprinkler system that the watering hose is connected to.

Watering Tube

The watering tube runs up from the pump in the reservoir and is sprinkled out through a hose bar that is attached to the end of the pipe. The hose bar is usually a bar shape, so the water is pushed out on both sides and the nutrient solution is dispersed evenly through the tower.

Pots

The pots used are flexible net pots that will not break when manipulated into the grow tube opening. The flexible pot is necessary as you will want to be able to easily access the net pots without them breaking.

The pots are positioned so that they do not touch and are well spaced so that each of the plants receives enough water-nutrient solution.

Growing Medium

Grow mediums are a good idea to place in the netted pots as they give the plants extra support as well as providing better aeration and much-needed drainage. The grow mediums should not be too absorbent or hang onto moisture and must allow the root system to breathe.

Growing Medium that Works Well With a Vertical Garden is:

- Lightweight expanded clay

- Growstones

- Rock wool

Reservoir

The reservoir sits at the base of the grow tube tower, it is usually a tub upon which the grow tower fits into. The reservoir contains the water-nutrient solution that feeds the tower.

- The pump must be powerful enough to efficiently pump the water to the top of the tower. It is usually attached to a timer that works for the growing environment.

- The feeding tube or pipe runs from the pump up through the tower to connect to the hose tube for the sprinkler positioned at the top of the tower.

Plants to Grow in a Vertical Garden

Plants that tend to grow quite rapidly are best suited to a vertical garden.

These include:

- Basil

- Broccoli

- Cabbage

- Chard

- Chives

- Cilantro

- Cucumber

- Dill

- Eggplant

- Kale

- Lettuce

- Mint

- Mustard greens

- Peppers

- Spinach

- Strawberries

- Tomatoes

9. Types of Hydroponic System

There are many types of growth boxes / systems to help you with your Hydroponic gardening. You can either opt to grow your plant in a small bucket before moving to a bigger growth box, or go for a system where you can start to cultivate your crops on large scale. A growth system can either use a material such as Perlite or Rockwool, or use NO growth material at all. You can build two types of system based on the delivery method of nutrient solutions:

Passive system

Active system

PASSIVE SYSTEM

In a passive system, the plants' roots are in touch with the nutrient solution and the plants are supported using suspension. The principal disadvantage of this system is that it is difficult to support your plants as they grow. However, a passive system is a basic system and is easy for a beginner to set up. This system is portable and inexpensive. An example of a Passive System is:

Wick System

A wick system uses a lamp wick or wick made of nylon, polyester or rayon to supply nutrient solution to the roots. Commonly used growth material are vermiculite, Perlite or LECA.

A pot is supported above the nutrient tray solution and a wick soaked in nutrient solution is passed through the drainage hole into the nutrient tray. You must leave 10 cm of the wick inside the pot and ruffle the ends for better circulation of nutrient solution.

ACTIVE SYSTEM

Active systems are more efficient and productive; they use pumps to supply nutrient solutions to the plants and a gravity system to drain off excess solution, which is then reused. Various types of materials can be used to act as a quick drain system, such as Perlite, Rockwool, expanded clay pebbles, or coconut coir. If you are using coconut coir, a higher air holding ratio can be achieved by mixing equal volume of Perlite with the coir.

The principal difference between an active and passive system is that an active system uses pumps to supply nutrient solution whereas a passive system uses a wicking action to draw nutrient solution.

TYPES OF ACTIVE SYSTEM:

The following are the various types of active system:

Ebb and Flow system

Nutrient Film Technique

Drip System or Top Feed

Aeroponics

Dutch Bucket Method

Raft Method

Ebb and Flow

An Ebb and Flow System is also called a flood and drain system. Maintenance and set-up is not expensive and hence is a popular system.

Both long-term and short-term produce grow well in this type of system. Here, the nutrient solution can flood the material for 15 minutes every hour or two. Popular growth material is expanded clay pebbles, Perlite or Rockwool.

Ebb and Flow system can be automated using a computer. There is a uniform distribution of nutrient solution to all the plants in this type of system.

Nutrient Film Technique

This system uses an automated pump and reservoir system to supply and recycle nutrients. It is possible to grow more produce using this system. Plants are placed in an enclosed inverted 'V' shape channel, but as a result, plants can suffocate and die because of lack of oxygen.

Problems also arise if there is a power failure or equipment failure. This system is primarily used by lettuce growers. If you are a herb lover, use NFT to grow them; you will love the results!

Plant Tray Drain Pipe ReservoirW ater Pump Fill Pipe Nutrient Pump Reservoir Nutrient Return Airstone Air Pump

Drip System or Top Feed System:

One of the principal advantages of this system is that it can withstand short-term power / equipment failure. Rockwool is used as the material here. Nutrient solution is dripped onto the plants and the remaining solution is drained back to the reservoir. Supply of nutrient solution is timed. It is expensive and difficult to set up a drip system. However, it is popular among tomato and pepper growers.

Nutrient Pump Drip Manifold Reservoir Airstone Air Pump

Aeroponics

A recent development in which plants are suspended in midair and are supplied with nutrients. Nutrients are sprayed to the roots; their exposure to air provides them with maximum oxygen. In this system, supply of nutrients and oxygen is maximized. Care has to be taken to maintain 100% relative humidity. The principal drawback to this system is the functioning of pump and reservoir in the event of power failure. It is expensive to set up this system and is more often used in laboratory studies.

Dutch Bucket Method

This system was first used in the Netherlands to grow tomatoes, cucumbers and roses. This uses a bucket (2.5 gallon) that holds nutrient solution at the bottom of the bucket. A pump is used to recycle the nutrient solution.

Raft Method

In this method, Styrofoam sheets are used to float plants fixed in baskets on top of nutrient solution. Usually, short term crops are cultivated using this system. The problem of stagnation is solved by circulating air from bottom. This system is used for lettuce production and to cultivate other greens.

HOW TO BUILD YOUR OWN HYDROPONIC SYSTEM

You might want to build your own system. One of the main advantages of making your own system is that you can customize. It is also easy to build.

Here you have three alternatives to consider.

Deep water culture

In this DIY guide we will show you step by step how to construct a simple Deep water culture system for 6-12 plants. This type of system is very easy to construct even for a beginner.

MATERIALS NEEDED

Air pump 3+ watts

12 inch Air stone

air tubing a few feet will do

18 gallon tote(Rubbermaid) container with lid, this will be the reservoir

6-12 3 inch net pots

Clay pellets growing medium-1.5 liters per net pot

Rockwool cubes

Marker

Box cutter

Power Drill with ½ drill bit

Hydroponic nutrients

INSTRUCTIONS:

First thing you need to do is figure out where the net pots are going to be positioned. Use a marker to make some 3 inch circles where the net pots are going to go. After this is done start cutting the holes out with a box cutter.

Now that this is done set the lid aside and start washing out the tote, preferably with 5%-10% bleach to 90%-95% water mixture.

Now drill a small ½ inch hole at the top of the tote by the handle part, make sure it is close to the top so water won`t escape out. This hole is where the tubing from the air pump will go.

Now measure some tubing from where the air pump will be outside the system all the way through the hole you drilled and to the bottom of the tote. Now cut the piece of tubing and connect one end to the air pump and run through hole and connect air stone to other end and place air stone on bottom of tote.

Now you can fill with water. Try to fill well below where you drilled hole, you can add or subtract water later on if needed. Next add desired nutrients to water, follow the directions that came with the nutrients carefully. *optional, run pump overnight to evaporate chlorine is water source is known to contain chlorine.

Next install the net pots to the holes in lid, they should fit nice and snug. Fill the net pots with clay pellets. After this is done put the lid on the tote and make sure the bottoms of the net pots are submerged in the nutrient solution.

I recommend first starting plants in rockwool cubes then transplant into the net pots filled with clay pellets. But your starter medium of choice should be fine.

That's it; you now have a fully functional deep water culture hydroponic system.

Drip feed system

This how to guide will show you how to construct a simple 4 plant drip feed hydroponic system.

MATERIALS NEEDED

5-8 gallon square bucket with a lid

Small submergible water pump 105 GPH will do

air pump

6 inch Air stone

5 feet of ¼ inch drip tubing

3 "T" connectors for the drip tubing

3 feet of ¼ inch air tubing

Four 3 inch net pots

1 ½ pounds of clay pellets hydroponic medium

Box cutter

Drill for drilling holes

Hydroponic nutrients

INSTRUCTIONS

Start out by tracing four, 3 inch diameter circles on the lid where the net pots will go. Next using a box cutter carefully cut out the holes. Now drill a ¾ inch hole directly in the center of the lid, this be for the drip tubing. And now drill a 1 inch hole near the edge of the lid; this will be used for the power cord running form the water pump and air tubing coming from the air pump.

Next place the water pump at the bottom of bucket, running the power cord through the 1 inch hole, and connect one end of drip tubing to it. Run the tubing on the side of the inside of the bucket and through the middle hole, and cut the tubing once a little outside the hole on the lid. Connect one of the T connectors to this. Now measure and cut 2 pieces of tubing that will run from both ends of the T connector to the middle of where the 2 net pots on each side will go. Place the remaining 2 T connectors on each end. Now cut 2 small pieces of tubing for each side of the connectors, so that they are just long enough to reach the center of the net pots.

Now connect one end of air tubing to the air pump and run the tubing from the air pump through the 1 inch hole and to the bottom and cut. Now on this end connect to the air stone and place on bottom.

Insert net pots into cut holes and add growing medium, which will be clay pellets in this case. Fill container with water/hydroponic solution mix. Fill up until the bottom part of the net pots are fully submerged in the water.

Now all you have to do is plug everything in and add seedling to the net pots and you're done. I suggest starting seedlings in rockwool cubes or oasis cubes and then transplanting to the net pots with clay pellets.

Ebb and flow

An ebb and flow system is an easy system to construct and any type of water pump is optional and not needed, with is totally unique then other systems, while still achieve optimal results.

MATERIALS NEEDED

Two 15-20 liter containers, buckets or Rubbermaid tots

Growing medium: I suggest all clay pellets or a 50/50 mix of perlite and vermiculite

Large gravel rocks or a few liters of clay pellets

3 meters of flexible tubing, any tubing intended for irrigation will do

2 tubing joints

2 tubing grommets

1 table: a generally medium sized table will do, as long as it is big enough to accommodate the 2 buckets

hydroponic nutrients

Drill to make holes

Silicone

INSTRUCTIONS

First off you want to use a drill and make a hole on the side near the bottom of both of the buckets, making sure the diameter is the same as the inside of the grommets that you will be using to connect the tubing and joints. These holes should be around 3-5cm above the bottom on each bucket.

Now insert the grommets into the holes you`ve just made, make sure that it is a tight fit, if done correctly this should be somewhat water tight. I recommend apply silicone around the holes before inserting the grommets to make sure that it is 100% water tight. Let this sit until silicone is completely dry.

Now connect the 2 joint pieces to each end of the irrigation tubing and insert each end of the tubing into the grommets of each bucket.

Now connect the 2 joints to each end of the irrigation tubing and then insert each end into the grommets on each bucket. Now the 2 buckets should be connecting together by the tubing.

Now use the gravel rocks and fill one of the buckets just enough to cover the bottom hole, this is imperative so no perlite or vermiculite will clog up in your tubing. Now that this is done add your growing medium of choice to the same bucket and fill until it is about 6cm from the top.

At this point you can add your seedling in the growing medium. After this is done fill the other bucket with water and hydroponic nutrients to about 6cm or lower from the top.

Now you are done. Your bucket that you will be growing in should always remain on the table and to flood system place other bucket on table and water will be transferred to growing bucket. To drain place same bucket under the table, simple enough right? You should flood your system approximate 5 times a day for 20-30 minutes at a time, and then drain accordingly.

LIGHTING REQUIREMENTS FOR HYDROPONIC SYSTEM

Sunlight is essential for plants' growth. Sunlight provides the energy for growth, germination, flowering and photosynthesis. In Hydroponic gardening, sometimes natural light is absent or difficult to provide. To overcome this drawback, we provide light to the plants through artificial means using artificial lighting. Usually lighting is provided for 16 to 18 hours every day to optimize growth. Care must be taken to provide complete darkness for the remaining duration of 6 to 8 hours. Specific plants like roses have 'photoperiodism' – meaning, the flowers bloom depending on the length of daylight.

There are many types of lighting available:

Fluorescent tubes

Incandescent lights

High Intensity Discharge

Metal Halide

High Pressure Sodium

Fluorescent tubes are low wattage bulbs that emit low-temperature light. This lighting is most suitable for the first two weeks of a plant's life. Thereafter, the intensity and heat generated from fluorescent lighting will not suffice for plants' lighting requirements.

Incandescent light bulbs emit light and an equal amount of heat; they are expensive to operate and are not healthful for plants' growth. You may consider converting incandescent bulbs into growth light bulbs by coating the inside – but it seldom works!

HID – High Intensity Discharge lighting is the most economical way to provide lighting for your plants; it is also the safest way. You can find HID lighting in parking lots, playgrounds, and at places that need high efficiency at low cost.

Blue-white spectrum is best suited during the vegetative phase of a plant's growth; this is provided by **Metal Halide lighting.** This lighting helps the formation of strong leaves, stems and branches. If you are planning to have only one lighting system for your Hydroponic garden, this is the best choice.

Best Suited for: Roses, Zinnias, Marigolds, Chrysanthemums, and Geraniums

During the flowering and fruiting phase, red spectrum light is most favorable for plants. You can provide this lighting through **High-Pressure Sodium** lamps. This lighting is usually used with Metal Halide lighting. Herbs like dill and coriander grow well under this light, and this light is primarily used in commercial greenhouses.

When using HID lamps, the spacing must be 12 to 14 inches from the plants for 250W and 16 to 24 inches for 1000W lamps. Walls are usually painted with light reflective coatings to increase the diffusion of light. Some common wall treatments include:

Aluminum foil – provides 60% to 65% reflection

Yellow paint – provides 65% to 70% reflection

Mylar – provides 90% to 95% reflection

Gloss white paint – provides 70% to 75% reflection

Flat white paint – provides 75% to 80% reflection

You can use light movers with HID for best results. By using movers, you eliminate the need for plants to grow toward the light.

There are two types of movers – linear and circular.

Linear movers, as the name suggests, move in a linear pattern (back and forth). These are around 6 feet long and carry a single lamp – most beneficial for narrow and long grow areas.

Circular movers can carry one, two or three lamps and cover the growth area in circular pattern covering 10 x 10 foot area. These are suitable for growth areas that are wide and long or square.

MAINTENANCE

Every type of hydroponic system has to be clean once in a while. Do it about once a month. When it comes to properly cleaning out a hydroponic system it is simpler than it sounds. Just soak you reservoir in hot water and rinse, then mix 10% bleach and 90% water in a spray bottle and thoroughly spray complete inside and scrub. As for the drippers and irrigation parts just let these soak in a 10% bleach/water mix for 10 minutes then clean thoroughly.

10. Mediums, Nutrients and Lighting for The Hydroponic Garden

Now that you have an overview of the different hydroponic systems, it is time to look at the growing mediums and nutrients needed to enable a

bountiful harvest.

BEST GROWING MEDIUMS FOR EACH SYSTEM

First, consider which type of mediums you want to grow your plants in.

Many growers combine mediums to gain all the advantages of each one.

When we talk about mediums, we are referring to the contents of the growing tray or pot that the plant's roots stand in.

AGGREGATE MEDIUMS

This method is also referred to as the Aggregate Culture. Often, mediums are mixed to allow for a combination of needs. There may be a heavier medium

at the bottom, such as small stones to provide weight and drainage. The main medium must provide stability for the plant to stand in a plant pot or small hanging basket.

Mixed in with that there must be a medium that can aerate well, such as perlite. Deeper rooted plants, such as chicory and beets, will need a system that has medium to support the heavier roots. Much is the same with topheavy plants, such as squashes, zucchini, and even tomatoes.

It is a case of trial and error in your chosen system. What works for one gardener may not work for another. This book can guide you through the first

stages, while you gain such experience.

DRIP SYSTEM

A system that works well with many plants, it can also be compatible with many different types of medium. The Drip System fares better with larger plants that retain the moisture for longer. With this system, a slow draining type of medium is best. Be aware that if not properly maintained the liquid nutrients may clog the feed pipes!

You can lessen this problem by using a filter, particularly if you are recycling or recovering your water system. It is advisable to flush out the pipes with plain, fresh water and cleaning fluids, periodically, to stop this from

occurring.

Rockwool

Available as a slab, small block, or loose fiber, depending on the planter pots and trays you have. Keep the medium covered with a protective layer to discourage algae growth. Such growth flourishes in the light. It is best to set the drippers intermittently as this medium is prone to flooding if overirrigated.

Allow the roots to dry out a little in-between the water feeding process, which is also good for the flowering and fruiting process.

Clay Pellets

Excellent with the Drip System. Not prone to flooding because the pellets are so porous, so the roots cannot become saturated. Because of the spherical shape of the pellets, it is a good aerating method, as oxygen becomes trapped between the pellets. If growing vegetables, it is preferred to use a constant drip cycle during the daytime. However, if growing fruit or flowers, a timed on/off cycle of 15 minutes each, during the daytime, is better.

Vermiculite and Perlite

Originally vermiculite contained asbestos fibers which can be harmful to your health. Now, modern products contain no asbestos. However, be wary of using older versions. It is more lightweight than perlite but they are both porous, so water and air flows through them as well. Plus, they can both retain water for a while when the pump is off. Both are considered good for this system as they are porous and allow for a good airflow.

EBB AND FLOW

Because this system requires you to flood your feed trays, a heavier medium that will not float away is required. Avoid lightweight mediums. Your medium should have a good Water Holding Capacity (WHC), and High Air
Filled Porosity (AFD).

With Ebb and Flow, your crop can take advantage of a mix of mediums. For example, coir retains water well, and clay pellets are good at draining, so the two combined for each plant will allow both good drainage and retaining moisture. A recommendation would be approximately 20% coir and 80% clay, so the coir does not retain too much water. It also means that should you have a system failure, each plant has a small amount of water in storage.

Rockwool

Coir

A great substitute for peat moss because it is more sustainable. It does not rot easily as it contains natural allergens. It retains water like a sponge. However,

the coir needs to be completely decomposed, otherwise, it can rob the plant of nutrients as it decomposes. If the coir is grown near salt water, it can also retain the salt. It holds air well. It may have a high pH balance which can be too acidic for certain types of plants. It can be re-used so long as it is not contaminated with any pests or water bacteria. It can have a natural fungus which will help plants fight off fungi although not if it has been sterilized.

Clay Pellets

The greatest asset is the rich supply of oxygen that the pellets can hold, causing roots to grow at a fast rate. Ideally, soak the pellets before potting to allow for expansion. For the germination of seeds, you could crush the pellets a little to make the water retention more efficient. It will hold nutrients for longer because of its binding process. This can cause a whitish substance to grow on the pellets which are salt. To avoid any toxic build-up, it is best to flush it out with plain water periodically. They should never be allowed to dry out nor should they be constantly flooded. You need a fine balance. This is why they are ideal for the Ebb and Flow System.

Perlite

Some might be a little nervous to use this medium, as it is made up of volcanic glass shards. However, it is heated until it expands into popcorn sized balls or smaller. It is non-organic and is often added to soil to aerate it better. It is a porous medium, allowing for good drainage. It serves well as an aerator medium. It is usually best mixed with another medium if it is being used for stability because it is lightweight.

NUTRIENT FILM TECHNIQUE

Whilst most NFT systems do not require any medium, fruiting and flowering plants do not do so well with their roots constantly wet. It is better to hang the plant in a medium so the nutritional water can be soaked up in between the water flow. This gives the roots the opportunity to dry out a little. The best mediums in these circumstances would be ones that soak up well, such as

Rockwool

Coir

For other plants that do not require such a "dry-out," it is better to try and ensure the roots do not become tangled by using:

Hanging Baskets

This requires no mediums because the roots are suspended in a hanging basket, sitting in a tray, with holes for the baskets.

WATER CULTURE

This requires no mediums because the roots are suspended in water.

AEROPONICS

This requires no mediums because the roots are suspended in air.

WICK IRRIGATION

Coir

Perlite

Clay Pellets

LIQUID CULTURE

There are hydroponic systems that require no medium at all, which is known as Liquid Culture. The plant roots are suspended into the nutrient-rich water.

This includes the NFT, Aeroponics, and Water Culture systems. Smaller rooted plants, such as herbs, and fast-growing roots, such as lettuce, are ideal for Liquid Culture. Also, plants with shallow roots, such as strawberries.

NUTRIENTS FOR THE HYDROPONIC GARDEN

Since hydroponic systems don't use soil, the nutrients added to the water must

contain all the minerals plants need to thrive. The selection of the appropriate nutrient solution is one of the most important decisions of your hydroponic plan. Healthy plant growth depends on having the right balance of nutrients.

There are 16 essential elements that plants need. These elements are absorbed by the plant in different ways. Some get transferred to the plant through the roots, while others are taken in through the pores on the leaf. Carbon, Oxygen, and Hydrogen, three of the most necessary, are available in both air and water. These need to be monitored and balanced. One common problem in hydroponics is a lack of sufficient carbon dioxide.

The next big three elements, Nitrogen, Phosphorus, and Potassium, are provided in the fertilizer nutrient blends made for growing hydroponic plants.

A fine balance of these is extremely important. This is often referred to as the N-P-K mix. When looking at bottles or bags of fertilizer, you'll see a list of three numbers on the front, separated by dashes. It will look something like this: 3-4-1. These three numbers refer to the Nitrogen, Phosphorus, and Potassium (N-P-K) proportion of the mix.

Calcium, Magnesium, and Sulfur are the next most essential elements. They are also supplied by fertilizer supplements. Calcium is provided through a

calcium nitrate (CaNO3) fertilizer. Magnesium and Sulfur are available with magnesium sulfate (MgSO4) supplement.

The remaining 7 essential elements, Copper, Zinc, Boron, Molybdenum, Iron, Manganese, and Chlorine, are rarely deficient. If there is an Iron deficiency,

you can supplement your plants with chelated iron.

To make sure your plants get everything they need, specially crafted fertilizer mixes are made for hydroponic crops. These mixes can be added to the water in your reservoir and distributed to your plants through the hydroponic system. Specific fertilizers are created for specific crops. They're not all interchangeable. The hydroponic nutrient mix for tomatoes will be quite different from the one for lettuce.

Nutrient mixes are available as liquid or granules. Liquid fertilizer is easy to use. You just pour it into the water reservoir as per bottle instructions. The downside to using liquid fertilizer is that it is more expensive and bulkier to store. Granulated fertilizer is more cost effective, easier to store, and often comes in bulk. However, it isn't as easy to use because it has to be mixed prior to use and it doesn't always dissolve completely. Either one will work fine, so it's a matter of personal preference and what's required by your

particular system.

Granulated mixes are available in three types. The one-part mixes are simple and straight-forward. The fertilizer is mixed as indicated on the bag. These

are simple to use, but not the best for making stock solutions. Some nutrients in high concentration will form solids. A multi-part solution is better for making stock solutions because the compounds are kept separately. They are relatively easy to mix, too. This is the most common choice for growers who are using granulated nutrient mixes. However, the multi-part mix can be expensive, so it isn't the best choice unless you have a very large garden operation.

Hydroponic fertilizer mixes are also specialized for different stages of growth. They will indicate on the package the stage of growth for which they are designed. Examples of growth stages include vegetative (leaf growth) or blooming (flowering). You'll want to know what you actually want from the plant. Spinach, lettuce, and kale, for example, will benefit most from

vegetative growth because you harvest the leaves of the plant. For plants that deliver a fruit or flower, you'll want to use the vegetative mix up to the point

where you want them to flower. Then, switch to the flowering mix. Nutrients are further classified based on the growing medium that is being used. The majority of nutrient mixes are made for a specific growing medium. Pay attention to package specifics and do your homework. If you're using a vermiculite mix but your growth medium is clay pellets, then you won't get optimal results. If the package doesn't give you all the details you need, then a simple online search will lead you in the right direction.

Hydroponic nutrients can be organic or synthetic. Organic fertilizers are best for systems that recirculate or reuse the nutrient solution. These mixes often

include materials that can clog up sprayers, drip lines, and pumps. Synthetic nutrients don't have this issue and are therefore more commonly used in hydroponic systems. Organic fertilizers will often have a lower N-P-K listing than synthetic. However, this doesn't mean they are lower quality. Synthetic mixes are generally fast-release, as opposed to the slow-release of organic, and so the readily available N-P-K is higher in the synthetic. However, organic mixes will deliver a natural, time release fertilizer that won't burn your roots.

NUTRIENTS YOU NEED FOR YOUR HYDROPONIC SYSTEM

An N-P-K mix, formulated for the crop you are growing
Calcium Nitrate (CaNO3)
Magnesium Sulfate (MgSO4)

ADDING NUTRIENTS TO YOUR HYDROPONIC SYSTEM

1. Mix the solution as per the package instructions and add it to your
reservoir.
2. Check your pH balance on a daily or weekly basis. (Timing will depend on the system you are using and the crops you are growing.)
3. Change out or top off your solution weekly or bi-weekly. (Again, the timing depends on the system and crops).
4. Flush your crop before harvesting. (Flushing your hydroponic crop
means allowing it to grow without nutrient solution for a brief period prior to harvesting.)

WHAT TO WATCH OUT FOR

Even with a premix nutrient, issues can still arise. This is often due to incorrect strength, usually because of over dilution. Other problems may occur when you use a dry mix and dissolve it in the water yourself. The result

of an incorrect balance will be a mineral deficiency in your plants. Follow the manufactures recommendations carefully using an accurate tool for the measurements.

Signs of mineral deficiency could include:

Yellowing of leaves

Stunted growth

Wilted or blackened leaves

Swollen or discolored root tips.

Whichever option you prefer, the solution must maintain a constant temperature within the range of 70-80F/21-26C. This is to ensure optimal growth for your plants.

TESTING THE PH

pH is the balance of acidity and alkalinity in your water. The nutrients you add to your water will influence the acid/alkaline balance. pH is measured on a scale from 0-14, with 0 being the most acidic and 14 being the most

alkaline or basic. pH tests will tell you how well the plants will be able to use the nutrients. Each plant prefers a specific pH balance, and your plants won't

be able to absorb the needed nutrients if the pH is too high or too low.

Measure the pH after you've added the nutrients and then adjust as needed. A good baseline is to keep it between 5.5 and 6.5.

pH testing devices include paper litmus test strips, liquid test kits, and electronic testing pens. Paper test strips are the cheapest way to go, but they lack accuracy. Litmus strips change color when you dip them in solution.

The resultant color reflects the pH. But checking the resulting color against the chart is a bit subjective, so you won't be able to determine the acidity of

the solution with any amount of precision. The results can also be skewed if your nutrient solution isn't clear, which is a problem because many nutrient solutions will color the water.

Liquid test kits offer a fair balance of cost and precision. To use a liquid test kit, you take a small sample of solution and place it into a vial which contains

a pH-sensitive dye. As with the litmus test, you will compare the resulting color with a chart. This will help you to determine the pH balance of your solution. The color changes are easier to see, and the test is a bit more sensitive than a litmus test, so the liquid test kit is a bit more accurate.

However, liquid test kits can also be skewed by the color of your solution (if it's not clear), so they aren't 100% accurate. However, unless your plants are

extremely sensitive, liquid test kits are accurate enough.

If your number one consideration is accuracy in pH testing, digital meters are the way to go. They are more expensive, but they will tell you the pH to a tenth, and they won't be skewed by the color of your nutrient solution. To use a digital meter, you just insert the tip of the meter in the solution and it will provide you a digital reading. The one thing you have to watch out for with these meters is calibration. To calibrate them, you must dip them in a pH neutral solution to provide a baseline. This is easier than it sounds, and you can find plenty of information about it online if you need.

If you need to adjust the pH, phosphoric acid will raise acidity (lower pH) and lemon juice will lower acidity (raise pH). There are also a number of pH adjustment products readily available in hydroponic stores.

FLUSHING

The nutrients you feed your plants build up in them and can cause bitter or chemical tastes. Flushing out the plants before harvesting ensures a good end

product. Do this for 4-7 days prior to the harvest. The most traditional way to do this is to irrigate your plants with pure water and allow them to process it through their system for up to a week before harvesting. If you'd like to get fancy, several flushing agents are readily available at hydroponic stores.

They'll speed up the process and ensure a complete flush. Remember – flushing is extremely important. You'll be able to taste the difference, regardless of what it is you're growing.

OPTIONAL ADDITIVES FOR THE PLANTS

BLOOM MAXIMIZERS

These are added to your nutrient solution to increase the size and yield of your plants. They are usually high in Phosphorus and Potassium. This

additive can be a bit expensive, but it's generally worth the price for the boost

it gives the plants. Nutrient burn can be a problem when using this so monitor the plants closely if you choose to use it. (Nutrient burn is the plant's

equivalent to chemical burn. If you see the roots turn colors, take on an unhealthy texture, or shrivel after adding the solution, flush the reservoir with pure water so that the plants can recover. It's far better, though, to make sure that you use the right concentration and you don't have to resort to damage control.) Bloom maximizers should only be used during the flowering stage of growth.

MYCORRHIZAE AND OTHER FUNGI

Mycorrhizae are small fungal filaments that penetrate the roots, increasing their surface area. They also gather and break down certain nutrients.

Mycorrhizae exist in a symbiotic relationship with nearly all plant species.

They help plants to absorb nutrients and water. In return, they receive some of the sugars that plants create through photosynthesis.

Mycorrhizal fungi can be added directly to the nutrient solution and will grow alongside the roots as they do in nature. You can also add other fungi like Trichoderma to aid in breaking down nutrients and making your crops more resistant to soil pests. Trichoderma and Mycorrhizal fungi are readily available in hydroponics stores. They will help your plants to remain healthy and grow more quickly .

VITAMINS AND ENZYMES

Thiamine (vitamin B-1) supports and strengthens the immune system of plants so they can better withstand stress and disease. It also facilitates root development, making the plants more resistant to shock and helping them to take in nutrients more quickly. This is especially important when transplanting. Enzymes break down nutrients, making them easier for plants to absorb. They are also helpful for preventing algae growth.

ROOT STIMULATORS

Root stimulators are compounds that replicate the benefits of natural soil.

There are beneficial microbes in soil that promote plant growth, just as there are harmful microbes that interfere with plant growth. Rooting stimulators introduce the healthy microbes into your hydroponics system, helping your plants to have stronger immune systems, more access to nitrogen, and faster root development. They are also excellent at preventing bacterial complications in the root structure.

Overall, root stimulators promote fast, healthy plant growth. If you add root stimulators at the beginning of your growing cycle, they will continue to reproduce throughout your plants growth from seedling to harvest, providing more robust, faster-growing crops from the start.

NUTRIENT NOTES

When searching for nutrients, you will encounter a slew of brands and products. While they'll all claim to be the best, there's a great deal of variety in quality from one brand to the next, even among products designed for the same purpose. A brand or company might be good for one thing, but not so great for another.

The best way to deal with this is to read reviews from several growers to find out which products they prefer. This will provide you solid feedback from people who have been there. Find hydroponic forums where you can post the details of your system and crop. You'll get plenty of responses from experienced growers that will direct you to products that have worked for them in similar situations.

Water quality is of utmost importance in a hydroponic system. Do not underestimate the necessity of good clean water. Distilled or RO (reverse osmosis) water is the best choice. Tap water or city water can have pollutants, chemicals, additives, and any number of things that can potentially have a negative impact on plant growth. This being said, plants use *lots* of water. If

your prime concern is economy, then you'll use what you've got. Just remember that you get out of the plant what you put into it.

TYPES OF LIGHTING

Plants needs around twelve hours of light per day. Of course, this will vary depending on the plant that you are growing. Some plants prefer a great deal of light, while others do quite well with only a moderate amount. Remember that plants get their energy from light. If your hydroponic system isn't in a place where it is getting natural light from the sun, you'll need to set up a lighting system.

Plants have rhythms, just like we do. Look into the preferred light cycles of your plants, and set up timers so that you give them a schedule as close as possible to their natural cycle. The optimal light schedule will differ

depending on the growth stage of the plant as well. Many plants grow well vegetatively when provided with constant light, but need cycles of light and darkness to trigger flowering.

The type of lighting you need depends on a wide variety of factors specific to your system: enclosure type, plant type, system size, ventilation, and last but not least, budget. Fluorescent tubes are good for a single low-budget system.

Small systems will fare better with CFLs (Compact Fluorescent Lamps).

These lights were designed as an efficient alternative to incandescent bulbs.

They screw into a standard socket and provide ufficient light, but you may want to arrange reflectors so that the light is focused on the plant.

HIDs (High Intensity Discharge lamps) are another option. They are a bit more costly than CFLs, but they are a preferred lighting option for experienced growers. This is because they have a very high light output and

are from four to eight times more efficient than standard incandescent bulbs.

However, they produce a lot of heat, so you'll have to ventilate your system to prevent it from drying out.

Another option is to use LEDs (Light Emitting Diode lamps). This is the high-tech option and will cost quite a bit more at the beginning, but they use a fraction of the electricity of other options and produce less heat. LEDs can also be calibrated to produce the exact spectrum of light that your plant needs. If you only plan to grow one crop, it's probably not worth it to purchase LEDs. But, if this is the beginning of a long relationship with hydroponic growing, they will more than pay off in the long run.

FLUORESCENT LIGHTS

Fluorescent lights are available in a wide range of sizes and spectrums. They are not ideal for large plants but they will work.

They are generally inexpensive, easy to set up, and will work in a pinch.

COMPACT FLUORESCENT BULBS

These are a good choice because they aren't too expensive and they don't require any special wiring or set-up (they screw into a regular light socket).

They produce light in all directions and are best used with a reflector so you don't waste any heat or light.

HIGH-INTENSITY DISCHARGE (HID) LIGHTS

HID lights have a high light output and are a preferred choice for growers.

Using HIDs will be better for your plants than fluorescents and less expensive than incandescent bulbs in the long run. At the same time, they are

expensive, require a special set-up, and need ventilation because of their high heat output.

LED LIGHTS

LED lights can provide the exact spectrum of light your plant needs. They last longer than all other lights and use less electricity. The upfront cost, however, is imposing. As mentioned above, LEDs are worth it if you plant to be growing many crops over the years. For a once-off, though, you may be better off with a different option.

11. Common Problems and Troubleshooting

Every system even a well-run hydroponic system can have a few problems whether technical or with the plants. This can be anything from algae build-up to a faulty switch on a pump.

Here are a few common problems to consider:

Nutrient Deficiencies

One of the most important aspects of watering plants in a hydroponic environment is that the nutrient levels must be just the right balance for the plants they are nourishing. An imbalance can cause a number of problems for the plants.

To ensure that the plants are getting what they need, an advanced nutrient solution of good quality should be used. Make sure that the pH levels stay in the range of 5.8 to 6.3 and the nutrient EC levels are not lower than 1.2 or higher than 2.0.

When plants are in the soil, they find the nutrients they need and do this by extending their roots out to find a mineral source. In hydroponic gardens, there is no soil and the roots only support system is a non-nutritious growing medium. They rely solely on the nutrient solution mix that is being fed directly to their roots.

Although the roots may be getting fed various nutrients in a well-mixed water-nutrient solution, it may not be the right solution for that particular plant. In a mixed plant environment, this can happen because what one plant type is happy with another may not be. Just like how one person may need a little extra of this vitamin while another is okay without. The good thing about most of the nutrient solutions for hydroponics is that you can find one that will suit all your plants. The trick is to know what to look for in your current plants to determine what it is they may be lacking. There are a few ways to diagnose certain nutrient deficiencies:

Boron Deficiencies

- The younger leaves are the first to be affected by a boron deficiency.

- The young leaves of the plant start to turn yellow.

- A boron deficiency will stunt the plant's growth.

- Any flowering buds will die without opening.

- The roots of the infected plant will be stunted and deformed.

- The overall appearance of the plant looks like it has been scorched.

Check the pH and EC levels of the nutrient solution. If they are fine and there are other like plants not affected it may be the positioning of the plant. Move it to a more prominent position and then try adjusting the nutrients to ensure that all plants are getting the correct nutrient intake.

If it is not the positioning of the plant, try changing the nutrient solution.

Calcium Deficiency

- Younger leaves are affected first.

- The new leaf tip, leaf edges, and inner parts of the leaf have brown patches on them.
- The leave tips are a bit distorted.
- The stems will have some black patches on them.
- The young leaf will die off and not last very long.
- Excessive buildup of calcium in a plant will stunt the plant's growth.
- The plant will have problems taking potassium and magnesium which in turn will cause more problems.

The only way to counteract this deficiency is to adjust the nutrient solution or use one with advanced nutrients in it. Keep the pH and EC levels at an optimum and remove any affected leaves.

Cobalt Deficiencies

- Cobalt is needed in plants as it is the bridge that allows them to take in other nutrients such as nitrogen and various metals.
- The younger leaves get affected first and will appear light green, almost a pale yellow, due to chlorosis.

This is a very rare deficiency but if it does happen, check the tank for various algae build up. It may be time to refresh the reservoir water. Try a new advanced nutrient solution and ensure that the water to nutrient ratio is correct.

Copper Deficiency

- The younger leaves will start to turn beige as chlorosis sets in.
- Leaf tips will curl and be malformed.

- The plant will suffer from an irregular growth rate.
- Too much copper is just as poisonous to a plant and if this happens, check for acidity in the nutrient solution.

Clean up the toxicity levels in the nutrient solution and the nutrient solution levels should be adjusted.

Iron Deficiency

- This is one of the most common deficiencies in plants and shows up a lot in plants that are grown indoors.
- Young leaves have darker leaf veins while the middle of the leaf has a yellow tinge to it.
- Younger leaves are a lot smaller than the older ones.
- Iron problems tend to occur when the water solution starts to turn acidic.

Check the pH levels and either clean out the tank or add more clean water to bring the levels back to normal.

Magnesium Deficiency

- Older leaves will get affected first.
- The outline of the leaf starts to turn dark and the discoloration will make its way towards the middle of the leaf.
- Discoloration starts to turn the leaf white in places, and it will eventually die.

The only way to counteract a magnesium deficiency in a hydroponic system is to augment the nutrients with an advanced calcium/magnesium solution.

Manganese Deficiency

- Young leaves will get affected first.

- The young leaves start to take on a wrinkled effect with mottled veins.
- Discoloration starts to turn the leaf white in places, and it will eventually die.

The deficiency of manganese comes from the root zone usually because the pH levels in the water nutrient solution have risen to well over 6.8. Bring the pH levels down and clear off the dead leaves.

Nitrogen Deficiency

- Older leaves will get affected first.
- The older leaves will start to take on a wrinkled effect with mottled veins.
- Discoloration starts to turn the leaves yellow and it will eventually die.
- Older plants can naturally lose nitrogen when near the end of their life span.
- If nitrogen is in excess, it will turn the leaves a darker green than normal. Excessive nitrogen affects the plant's ability to absorb nutrients and will dehydrate them.

Any form of nitrogen upset will cause plant harm. Check the pH levels of the reservoir are within the optimum range and used advanced nutrient products to counteract either a deficiency or excess.

Phosphorus Deficiency

- Older leaves will get affected first.
- The older leaf will start curl backward.

- The plant's growth rate will slow down and they will become spindly.

To adjust phosphorus levels, you will have to use advanced nutrients. Check the pH balance of the nutrient solution as well as the EC levels. The tank may need a complete cleanout and fresh solution.

Potassium Deficiency

- Older leaves will get affected first.
- The older leaf will start to go yellow and brown giving it a scorched look.
- If the plant is flowering, buds will fall off prematurely and die.
- Plants are stunted and the growth rate slows down.

The only way to augment potassium deficiency is with an advanced nutrient solution added to the nutrient mix.

Infestations

Infestations happen by way of pests and diseases. These are especially common on indoor grown plants.

Common Diseases:

- Algae

This usually grows on the growing media of the plant but can start to creep up on the plant itself. It is a thick green colored gooey mess. Although not a great threat it is still an indication that there is something not right with your growing environment.

Clean out all vents, change growing media and clean pumps, filters, etc. Change the reservoir water.

- Damping Off

This disease mainly affects seedlings where their growing media is far to damp. Damping off is fatal to the new plant and there is no way to cure it. Once it has attacked the seedling the plant will die. The root of the infected seedling looks like it is waterlogged. You can prevent the disease by making sure that the media used to house the seedlings is fast draining and sterile. Make sure the seedling cubes are not kept too moist and they are properly drained.

- Downy Mildew

This is not to get confused with powdery mildew. It is a white substance that appears on the upper surface of the leaf, making it look like it has been burned by a cigarette.

Mildew is caused by damp conditions from the humidity in a growing area especially an indoor one. It can usually be wiped away, although it is best to get a non-toxic cleanser from a local nursery to clean the spores away. To discourage mildew, clean away dead leaves, old flowers, produce, etc.

- Gray Mold

This looks like fine cobwebs or silvery hair that usually extends from the leaves of the plant. It is quite common on tomatoes and is caused by too much humidity. Gray mold is not an outbreak you want, once it takes hold it can be quite devastating to the growing environments. and should be caught as early as possible. Wipe away all the mold that can be found with a soft cloth, making sure to remove every bit of it. Remove all fallen leaves, old flowers, discolored leaves, etc. Each plant in the grow room will have to be thoroughly checked to ensure there is no mold on it.

The entire grow room should be thoroughly cleaned. The humidity in the room will need to be decreased and the air inflow increased.

- Powdery Mildew

Powdery mildew looks just like it sounds as if spots of powder have been dropped on the plant's leaves.

It is caused by high humidity, dampness, and not enough light.

The way to treat it is to decrease the humidity, increase the air circulation in the room and clear off the infected leaves. Make sure there is sufficient lighting for the plants. If there is too much, a fungicide may have to be considered.

- Root Rot

Root rot is when the roots get infected by a pathogen that chokes the roots and makes them become a slimy mess. Because the roots are not performing properly, the plant's growth slows and becomes stunted.

The plants themselves will start to turn yellow as they will be starved of nutrients. As the roots start to rot and decay, the plant will wither and die.

Change out the nutrient solution and clean the tank reservoir, pump and pump tubing. Completely clean out the hydroponic system and check all the plants in the infected grow area. Plants that are dying should be taken out of the rooms and those that can be saved must have their roots thoroughly drained and any dead or rotting roots removed.

The only way to clear root rot is to completely flush the entire hydroponic system.

- Wilts

Wilts affect plants like tomatoes, eggplants, and peppers. It causes small spots on the plant's leaves which eventually makes them start to curl up and dry out.

You will notice the leaves of the produce have browned, hardened, curled and dried out.

The only way to cure this disease is to get rid of the infected plants and completely flush out the hydroponic growing environment. Clean out all the dead plants and any debris around the room.

Common Pests:

- Aphids

These are commonly known as plant lice. They are either green, grey, or brown in color. They suck all the sap out of the plant's leaf and stem causing it severe damage. An attack of these critters will first show as the leaves start to turn yellow. They like to gather in a colony at the stem of the plant.

Aphids will need some form of organic pesticides to get rid of them. To prevent another attack, it may be beneficial to apply something like Rhino Skin.

- Fungus Gnats

The larvae of the fungus gnat are what attacks and destroy the plant. The adult gnat is not harmful to the plant. The larvae however can cause the plant's growth to slow, which can cause a bacterial infection and the death of the plant.

Sticky traps placed at the soil level of the plant will trap the gnats. The best cards for fungus gnats are the yellow ones. Once again, an organic trap is the best solution to kill off the larvae.

- Spider Mites

These are teeny little arachnids that are so small that they usually go unnoticed until they have caused major damage and increased their populations. The only sure way to spot if a plant has been infested with these mites is to look out for a fine webbing.

These are nasty little critters to try and get rid of but not impossible. It is going to take an organic pesticide. But the plant will have to be constantly monitored until all the eggs and any new hatchlings have been dealt with.

- Thrips

Thrips are tiny little insects that love to suck plants dry of their juice. Like the spider mite they are so tiny they are hard to spot. The only real way will be when the leaves on the infected plant start to get a metallic, black looking spot on the top of the leaf. Soon thereafter it will turn brown and wither up as the thrip sucks the leaf dry.

An organic pesticide will have to be used and the plants leaves wiped down. Any leaves that have the metallic spot must be plucked. Keep checking the rest of the plants in the grow room to ensure that all the thrips have been taken care of.

- Whiteflies

Whiteflies are really annoying and look like very small moths. Because they can fly, they tend to be quite hard to get rid of. Once you disturb a plant they are on, they will fly off to the next one. They breed and can spread at an alarmingly fast rate.

Once again organic pesticides and sticky traps are the keys to getting rid of these flies. Keep an eye on the plants and make sure to continue the application for a while after the first infestation has been taken care of. You do not want to miss any larvae that may have been left behind as this will quickly start another infestation

Seedling Problems

Like any new life, seeds are at their most vulnerable in the first few days to weeks of their life.

There are a few common seedling problems in hydroponic growing. In order to be able to respond quickly to them these little guys need you to check on them every day and keep an eye out for the following signs of a problem:

Seedlings do not Grow

- Most fresh seeds will grow without a problem. If a seed has been stored for longer than six months or so there is a good chance that it may not grow. This is because the older a seed gets, the less germination rate it has.

If a fresh seed, however, does not germinate there may be a problem with the growing medium or nutrient solution.

The best way to get a seed that has low germination to grow is to soak it for half a minute in a bit of water. This will soften the seed and help it to absorb moisture and encourage it to grow. Once the seed has been softened put into a grow block and make sure it has enough water but not too much as to drown it.

- The seedling has a spindly stem and the leaves are too small

This is due to seedlings trying to stretch towards the light and means that they are not getting enough direct light.

If you are using artificial grow lights, try to place the seedlings closer to the light. But be careful not to place them too close to the light as that will be just as damaging for the tender plants.

- The stems and leaves are all droopy

This is a classic sign of overwatering. The plant needs to dry out a little and the roots need some air.

Make sure the pot is not too big for the plant and make sure the growing medium is not holding too much moisture for the little plant.

- The leaves have yellow streaks and are pale

If, after a few days, the seedling suddenly stops growing and the leaves turned yellow and streaky, it means the little plant is lacking in nitrogen.

To begin correction a nitrogen deficiency in a seedling you will have to wait for it to produce at least two cotyledon leaves. Once these appear mix some nitrogen-rich advanced seedling solution in with their nutrient mix. It must be properly diluted following the manufacturer's specification in order to correct the deficiency.

- Leaves are curled and drying up

The seedlings are not getting enough water and nutrients. Seedlings need to be watered constantly and even one missed watering session can cause them damage.

Another cause of curled dried up leaves could be that the seedlings have too much heat on them from something like a grow light. Or they are positioned where there is too much direct sunlight.

Keep the plants cool and in a growing medium that retains just enough moisture to ensure the seedling has a constant supply of moisture to keep them from dehydrating.

- Leaves are purple or red

Newly developing plants need a lot of nutrients and a lack of them will cause the leaves to start to discolor. When the leaves of a seedling turn purple or red, it means the plant is lacking phosphorus.

The nutrient solution pH should be tested, and a correct balance established. If the nutrient solution is not ideal or too acidic, the little plant will not be able to absorb the nutrients it desperately needs to grow and thrive.

- Leaf tips are brown or yellow

If the room temperature is too warm the seedling will get too hot and will start to lose moisture as they overheat.

An ideal temperature setting for seedlings is around 65 degrees Fahrenheit.

Common System and Environmental Problems

Algae

Algae is not something that can be avoided when there is water involved. Especially when there is water mixed with nutrients and lighting. Although algae in itself is not the biggest problem and is not that hard to get rid of, it is quite a sneaky substance and at times things can go amiss.

Algae attracts the unwanted attention of pests such as fungus gnats and other nasty critters and growths. There are ways to try and prevent algae buildup such as:

• Use darker materials for the reservoir and try to limit the nutrient's exposure to direct light.

• The holes that feed the nutrient and drainage pipe should be just big enough to fit them through.

• These pipes should be made from materials that do not expose the water traveling through them to direct light.

• Check the plant baskets and growing medium regularly for any algae growth.

• Try to cap off any holes that may expose the solution to direct light that could attract the unwanted attention of algae.

System Leaks

• Hydroponic systems are prone to some leaks, especially the high-pressure systems.

• The biggest source of most of these leaks is usually at join points such as stab fittings, t-joints, and sprinkler systems.

• This could be due to:

o The stab holes being slightly too big for the fitting.

o   The joints not being pushed incorrectly for the hose or cap fittings.

o   The fittings not being the correct size for the piece it is meant to join.

Systems getting Clogged

- Pumps can get clogged.
- Pipes can get clogged.
- Plant baskets can get clogged.
- Sprinkler or spray systems can get clogged.
- Some growing medium such as coco coir is known for clogging drainage systems.

Incorrect Humidity

Hydroponic systems often have problems with incorrect humidity. This is something that can go wrong at any time. A heater or cooling system could malfunction, or a spray timer could be set incorrectly.

There are many factors that could cause a humidity problem and like algae, humidity problems have a knock-on effect. If not caught in time they can lead to other problems:

- Low humidity can cause leaf burn.
- High humidity is the perfect environment for various nasty fungi to grow.
- It can cause glassiness, edema, and tip burn.

Most crops are comfortable at a 70 to 75 % humidity.

Issues with Various Gases

One of the most overlooked pieces of equipment in indoor plant growing is a carbon dioxide monitor. This is needed in order to keep the various gases at levels that are both safe and acceptable to humans and plants.

- For an optimum carbon dioxide enriched environment, the levels should be in the range of 600 to 1,200 ppm.
- It is very important to ensure that there is a carbon dioxide monitor and that it is always working in growing houses.
- High levels of carbon dioxide in a grow house is toxic to the grower and can cause a person to lose coordination, become dizzy and pass out. Carbon Dioxide levels in the hydroponic growing houses should never get to levels in rangers of 5000 ppm.
- Carbon Dioxide poisoning to plants mimics a few nutrient deficiencies which it can easily be mistaken for. These include curling and browning of the leaves, yellowing and chlorosis, stunted plant growth and necrosis.

Chemical issues

Although Hydroponics does not use any actual chemicals, the plants can still be affected by chemicals from the nutrients supplied. This usually ends up being something to do with the quality or composition of the nutrient solution used.

For instance, tomato plants are not as susceptible to salt build up as cucumbers are. If cucumbers are exposed to too much sodium their leaves will get a yellow band around them. This could also be caused by the use of the city water supply, as certain chemicals used to treat city water can affect nutrient solutions.

Keeping a good chemical balance is not the easiest of things to do when running a system that is reliant on a water soluble nutrient system. The best water to use is that of rain runoff into actual catchment tanks that have not had any type of chemical treatments.

General System Problems

Troubleshooting any system can be complex as there are many factors that can influence both the problem and solution, especially in a system that has so many moving parts that need to work together in balanced harmony to give the desired outcome. A well maintained and monitored system will ensure that most of the working parts function at their optimum. Although it is not always feasible it is always a good plan to have a backup of the main parts like a spare pump, filter, fittings, and pipes.

That way the system downtime is not as long and there is a good chance that there will be no damage to the plants.

12.   Conclusion

Hydroponics is an economical, environmentally friendly way to grow plants and produce without soil or pesticides. The plants grow faster and produce bigger yields while being completely GMO-free, making them a lot healthier to eat.

Not only does hydroponics allow for fast, efficient, cost effective growing environments, but it is a means to grow produce where it otherwise was not able to grow. Thanks to innovative irrigation systems and the use of various growing media, places that have inadequate soil composition are able to grow fresh produce.

Hydroponics also provides a growing solution for places that have little to no space for commercial growing lands. It has even been successfully tested in space. Hydroponics is not a new concept but has come a long way since ancient times and keeps moving forward in leaps and bounds with new methods being introduced along the way.

It is not a hard concept to grasp and some methods are really easy to learn. There are ready-made kits that one can buy and assemble for each type of system. But they are all capable of being homemade with materials found around the home.

Hydroponics is a great way to teach children the joy of gardening without the mess of dirt and as the plants grow relatively quickly it holds their attention better than normal gardening does.

There are many exciting growing opportunities to be had with hydroponics and if done right, you will be rewarded with bountiful, healthy crops.

Aquaponics adds another dynamic level to the sustainable green farming in that it utilizes natural nutrients generated from a fish tank to organically nourish a media bed. In turn, the media beds offer the fish tank clean water as they filter out all the waste products and return clean water to the fish tank.

Lightning Source UK Ltd.
Milton Keynes UK
UKHW020700140521
383712UK00006B/112